THE MONTE CARLO RAY-TRACE METHOD IN RADIATION HEAT TRANSFER AND APPLIED OPTICS

Wiley-ASME Press Series List

Introduction to Dynamics and Control in Mechanical Engineering Systems	To	March 2016
Fundamentals of Mechanical Vibrations	Cai	May 2016
Nonlinear Regression Modeling for Engineering Applications	Rhinehart	September 2016
Modeling, Model Validation, and Enabling Design of Experiments Stress in ASME Pressure Vessels	Jawad	November 2016
Combined Cooling, Heating, and Power Systems	Shi	January 2017
The monte Carlo Ray-trace Method in Radiation heat Transfer and Applied optics	Mahan	December 2018

C'est en forgeant qu'on devient forgeron.

THE MONTE CARLO RAY-TRACE METHOD IN RADIATION HEAT TRANSFER AND APPLIED OPTICS

J. Robert Mahan

Virginia Polytechnic Institute & State University
Blacksburg, Virginia
USA

This Work is a co-publication between ASME Press and John Wiley & Sons Ltd.

WILEY

ASME PRESS

This Work is a co-publication between John Wiley & Sons Ltd and ASME Press

Registered Offices
John Wiley & Sons, Inc., 111 River Street, Hoboken, NJ 07030, USA
John Wiley & Sons Ltd, The Atrium, Southern Gate, Chichester, West Sussex, PO19 8SQ, UK

Editorial Office
The Atrium, Southern Gate, Chichester, West Sussex, PO19 8SQ, UK

For details of our global editorial offices, customer services, and more information about Wiley products visit us at www.wiley.com.

Wiley also publishes its books in a variety of electronic formats and by print-on-demand. Some content that appears in standard print versions of this book may not be available in other formats.

Library of Congress Cataloging-in-Publication Data

Names: Mahan, J. R., author.
Title: The Monte Carlo ray-trace method in radiation heat transfer and applied optics / J. Robert Mahan.
Description: Hoboken, NJ : John Wiley & Sons, 2019. | Series: Wiley-ASME Press series | Includes bibliographical references and index. |
Identifiers: LCCN 2018036102 (print) | LCCN 2018043872 (ebook) | ISBN 9781119518525 (Adobe PDF) | ISBN 9781119518501 (ePub) | ISBN 9781119518518 (hardcover)
Subjects: LCSH: Heat–Transmission–Mathematical models. | Monte Carlo method. | Ray tracing algorithms.
Classification: LCC QC320 (ebook) | LCC QC320 .M358 2018 (print) | DDC 536/.201518282–dc23
LC record available at https://lccn.loc.gov/2018036102

Cover design: Wiley
Cover image: © Fig. 5.21, Nelson, E. L., J. R. Mahan, L. D. Birckelbaw, J. A. Turk, D. A. Wardwell, and C. E. Hange
Frontispiece, Le Forgeron, © 2018 Sylvie Barbi, used with permission

Set in 10/12pt TimesLTStd by SPi Global, Chennai, India

Printed in Singapore by C.O.S. Printers Pte Ltd

10 9 8 7 6 5 4 3 2 1

To Kory J. Priestley

Contents

Series Preface

The Wiley-ASME Press Series in Mechanical Engineering brings together two established leaders in mechanical engineering publishing to deliver high-quality, peer-reviewed books covering topics of current interest to engineers and researchers worldwide. The series publishes across the breadth of mechanical engineering, comprising research, design and development, and manufacturing. It includes monographs, references, and course texts. Prospective topics include emerging and advanced technologies in engineering design, computer-aided design, energy conversion and resources, heat transfer, manufacturing and processing, systems and devices, renewable energy, robotics, and biotechnology.

Preface

This book is a stand-alone treatment of the Monte Carlo ray-trace (MCRT) method as it is currently practiced in the field of radiation heat transfer. While intended primarily as a textbook for use by first-year graduate students in curricula such as mechanical and aerospace engineering, the suitability of the MCRT method as an optical modeling tool makes the content equally well suited to the needs of students and practitioners of applied optics.

Max Planck, in his seminal 1912 book **The Theory of Heat Radiation**, writes that when undertaking radiation heat transfer analysis

> *… it will be assumed that the linear dimensions of all parts of space considered, as well as the radii of curvature of all surfaces under consideration, are large compared with the wave lengths of the rays considered. With this assumption we may, without appreciable error, entirely neglect the influence of diffraction caused by the bounding surfaces, and everywhere apply the ordinary laws of reflection and refraction of light.*

In other words, Planck is alerting the reader that radiation heat transfer analysis is to be based on the principles of **geometrical** optics rather than on the more complex principles of **physical** optics. The material presented in the current book extends radiation heat transfer beyond this limited view. By including principles from physical optics we are able to attack problems inaccessible to geometrical optics alone.

During most of the century following the publication of Planck's book, the version of geometrical optics used in radiation heat transfer analysis has been based on its implications rather than on the literal application of its principles. Until the emergence of the high-speed digital computer after World War II, the ray-by-ray application of geometrical optics to complex geometries was simply not practical. By the dawn of the new millennium, however, rapid advances in computing power had made it possible to emit and trace a statistically significant number of rays as they were scattered, refracted, and eventually absorbed within complex enclosures consisting of thousands of surface and optically participating volume elements. In other words, accurate simulation began to displace approximate analysis as a means of describing radiation heat transfer. Today, virtually no serious radiation heat transfer calculations are performed using the antiquated "net exchange" formulation, which is based on the questionable assumptions of uniform surface heat flux and diffuse gray surfaces, and is incapable by itself of treating radiation in participating media.

The mathematical basis of ray tracing and the fundamentals of thermal radiation are presented in the first two chapters. This material prepares the ground for Chapter 3, in which the MCRT method is introduced and used to model radiant exchange among diffuse gray surfaces. After the completion of the first three chapters, the reader is already armed with the essential knowledge required to formulate realistic radiation heat transfer models for the wide range of applications typically encountered in industrial settings. The next three chapters extend the MCRT method to include radiant exchange among non-diffuse non-gray surfaces (Chapter 4), radiation in a participating medium (Chapter 5), and the treatment of polarization, diffraction, and interference in applied optics (Chapter 6). The additional theory required to support these latter topics is introduced as the need arises. The ease of transition from the basic material of Chapter 3 to the more advanced material of Chapters 4, 5, and 6 is remarkable. This is due to the inherent flexibility of the MCRT method itself, whose basic principles apply equally well to directional wavelength-dependent surface models as they do to diffuse gray surface models; and whose logical structure applies equally well to radiant exchange among volume elements as it does to radiant exchange among surface elements. The treatment of polarization, diffraction, and interference using the MCRT method follows naturally after the definition of a "ray" is augmented to include its wavelength, phase, and polarization state. Finally, Chapter 7 presents a formal statistical method for assessing the uncertainty, to a stated level of confidence, of results obtained using the MCRT method.

J. Robert Mahan
Blacksburg
March 2018

Acknowledgments

Most of what I know today, and much of the content of this book, I learned through my interaction with the 60 or so outstanding young men and women whose dissertation and thesis research I have had the privilege of directing over a long and rewarding career spent mostly at Virginia Tech. Without the intellectual stimulation they provided, this book simply would not have been possible. Over the years my research has been more or less continuously sponsored by the National Aeronautics and Space Administration, principally by the Climate Science Branch of the Science Directorate at NASA's Langley Research Center. Without this funding, there would have been no graduate students, and thus no book.

I owe an unrepayable debt of gratitude to my own professors at the University of Kentucky, especially to my infinitely patient advisor Clifford J. Cremers, but also to Richard C. Birkebak, John H. Lienhard, IV, and Roger Eichhorn. Their high standards of scholarship, integrity, and achievement have served as a lasting guide for me as I pursued my own academic career.

Finally, my wife, Bea, who has been my constant companion for more than 50 years, has made the voyage worthwhile.

J. Robert Mahan
Blacksburg

About the Companion Website

This book is accompanied by a companion website:

www.wiley.com/go/mahan/monte-carlo

There is a website for Instructor and Students separately. The Website includes:

1) Solution manual
2) Data Tables
3) Analytical tools
4) Matlab codes

Scan this QR code to visit the companion website.

1

Fundamentals of Ray Tracing

A ray is defined as the collection of straight-line path segments followed by an energy bundle from its point of emission to its point of absorption. The definition includes the possibility of intermediate reflection, scattering, refraction, and even diffraction events. Ray tracing involves the application of basic mathematics to the process of identifying the intersection of ray segments with surfaces. Most engineering and science students acquire the required mathematical tools long before they enter university. The current chapter provides a review of the mathematical principles governing ray tracing and the related issues of meshing and indexing.

1.1 Rays and Ray Segments

A **ray** is defined here as the continuous sequence of straight-line paths connecting a point on one surface, from which an energy bundle is emitted, to a point on a second surface – or perhaps even on the same surface – where it is ultimately absorbed. One or several reflections from intervening surfaces may occur between emission and absorption of the energy bundle. The path followed by the energy bundle between

---

The Monte Carlo Ray-Trace Method in Radiation Heat Transfer and Applied Optics, First Edition. J. Robert Mahan.

This Work is a co-publication between John Wiley & Sons Ltd and ASME Press.
Companion website: www.wiley.com/go/mahan/monte-carlo

reflections is referred to as a **ray segment**. Two situations are normally considered: either (i) the power of the emitted energy bundle does not change as it is reflected from one surface to the next until it reaches the surface where all its power is ultimately absorbed; or (ii) a fraction of the emitted power is left behind with each reflection until the remaining power is deemed to have dropped below a threshold value, at which point the ray trace is terminated. Both approaches have their adherents and are in common use, and both are developed in detail in this book.

1.2 The Enclosure

The **enclosure** is an essential concept in all approaches to radiation heat transfer analysis. We define the enclosure as an ensemble of surfaces, both real and imaginary, arranged in such a manner that a ray emitted into the interior of the enclosure cannot escape. Energy is conserved within the enclosure under this definition. If a ray does leave the enclosure through an opening, represented by an imaginary surface, the energy it carries is deducted from the overall energy balance.

1.3 Mathematical Preliminaries

Consider two points, P_0 and P_1, in three-dimensional space, as illustrated in Figure 1.1. Let the Cartesian coordinates of points P_0 and P_1 be (x_0, y_0, z_0) and (x_1, y_1, z_1), respectively. Then the vector directed from P_0 to P_1 is

$$\boldsymbol{V} = (x_1 - x_0)\,\boldsymbol{i} + (y_1 - y_0)\boldsymbol{j} + (z_1 - z_0)\,\boldsymbol{k}, \tag{1.1}$$

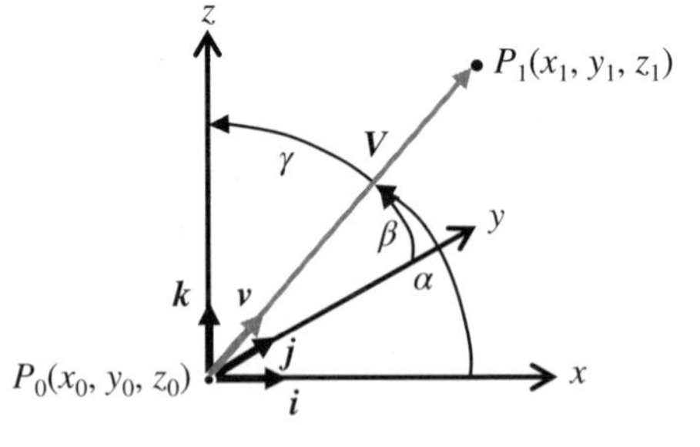

Figure 1.1 Relationships among the quantities introduced in Section 1.3.

and its magnitude is

$$t \equiv \sqrt{|\boldsymbol{V}\cdot\boldsymbol{V}|} = \sqrt{(x_1 - x_0)^2 + (y_1 - y_0)^2 + (z_1 - z_0)^2}. \tag{1.2}$$

In Eq. (1.1) $\boldsymbol{i}, \boldsymbol{j}$, and $\boldsymbol{k}$ are the unit vectors directed along the x-, y-, and z-axes, respectively. Note that the distance t from P_0 to P_1 must always be real and positive.

The unit vector in the direction of $\boldsymbol{V}$ is

$$\boldsymbol{v} \equiv \boldsymbol{V}/t = L\boldsymbol{i} + M\boldsymbol{j} + N\boldsymbol{k}, \tag{1.3}$$

where L, M, and N are the **direction cosines** illustrated in Figure 1.1. The direction cosines are defined

$$L \equiv \boldsymbol{v}\cdot\boldsymbol{i} = \cos\alpha, M \equiv \boldsymbol{v}\cdot\boldsymbol{j} = \cos\beta, \text{and } N \equiv \boldsymbol{v}\cdot\boldsymbol{k} = \cos\gamma, \tag{1.4}$$

where α, β, and γ are the angles between the unit vector $\boldsymbol{v}$ and the x-, y-, and z-axes, respectively. Equations (1.1) and (1.3) can be combined to define the equations for the line segment connecting point P_0 to point P_1

$$(x_1 - x_0)/L = (y_1 - y_0)/M = (z_1 - z_0)/N = t. \tag{1.5}$$

The three equations embodied in Eq. (1.5) are arguably the most important relationships in geometrical optics, because they form the basis for navigation of rays within an enclosure.

The general equation for a **surface** in Cartesian coordinates is

$$S(x, y, z) = 0. \tag{1.6}$$

The simplest, and perhaps most common, surface used in fabricating an enclosure is the plane, illustrated in Figure 1.2. In order to derive the equation for a plane, we must know the unit normal vector $\boldsymbol{n}$ at a point

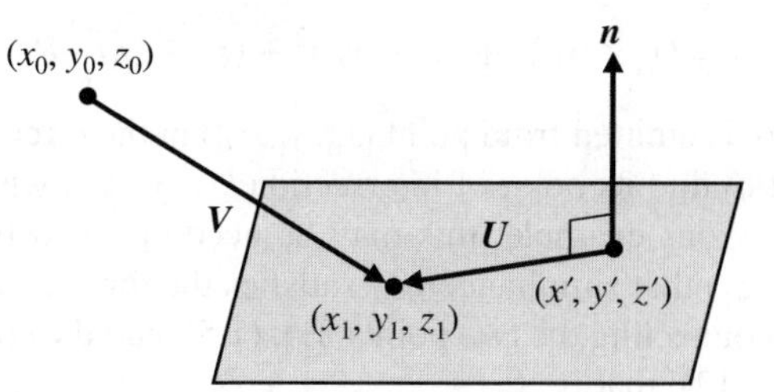

Figure 1.2 Definition of a plane surface.

(x', y', z') in the plane and the coordinates of a second point (x_1, y_1, z_1) in the plane. Then, because $\boldsymbol{n}$ and $\boldsymbol{U}$ are in quadrature, it must be true that

$$\begin{aligned} S(x_1, y_1, z_1) &= \boldsymbol{n} \cdot \boldsymbol{U} \\ &= \boldsymbol{n} \cdot [(x_1 - x')\,\boldsymbol{i} + (y_1 - y')\boldsymbol{j} + (z_1 - z')\,\boldsymbol{k}] \\ &= 0, \end{aligned} \tag{1.7}$$

or

$$S(x_1, y_1, z_1) = n_x\,(x_1 - x') + n_y\,(y_1 - y') + n_z\,(z_1 - z') = 0. \tag{1.8}$$

To find the intersection of the ray segment $\boldsymbol{V} = (x_1 - x_0)\,\boldsymbol{i} + (y_1 - y_0)\,\boldsymbol{j} + (z_1 - z_0)\,\boldsymbol{k}$ with the plane, we introduce Eq. (1.5) into Eq. (1.8), obtaining

$$n_x\,(x_0 + Lt - x') + n_y\,(y_0 + Mt - y') + n_z\,(z_0 + Nt - z') = 0. \tag{1.9}$$

Finally, solving Eq. (1.9) for t we obtain

$$t = \frac{n_x(x' - x_0) + n_y(y' - y_0) + n_z(z' - z_0)}{n_x L + n_y M + n_z N}, \tag{1.10}$$

or

$$t = \frac{\boldsymbol{n} \cdot (\boldsymbol{V} - \boldsymbol{U})}{\boldsymbol{n} \cdot \boldsymbol{v}}. \tag{1.11}$$

Note that if $\boldsymbol{n}$ and $\mathbf{v}$ are in quadrature, $\boldsymbol{n} \cdot \boldsymbol{v} = 0$, in which case t is undefined. The interpretation is that the ray passes parallel to the plane and so can never intersect it. We must anticipate this eventuality when programming. This is perhaps an appropriate juncture to emphasize the natural compatibility of Cartesian coordinates with the vector nature of ray tracing.

A more instructive example is the intersection of a ray segment with a sphere of radius R whose center is located at (x_C, y_C, z_C); that is,

$$S(x_1, y_1, z_1) = (x_1 - x_C)^2 + (y_1 - y_C)^2 + (z_1 - z_C)^2 – R^2 = 0. \tag{1.12}$$

Suppose a ray is emitted from point (x_0, y_0, z_0) in the direction (L, M, N) and we want to find its point of intersection (x_1, y_1, z_1) with this sphere. As in the previous example, this may be accomplished by finding the point (x_1, y_1, z_1) that simultaneously satisfies the three equations for the straight line connecting the two points, Eq. (1.5), and the equation for the sphere, Eq. (1.12); that is,

$$(x_0 + Lt - x_C)^2 + (y_0 + Mt - y_C)^2 + (z_0 + Nt - z_C)^2 - R^2 = 0. \tag{1.13}$$

Happily, the solution of Eq. (1.13) for the distance t is just about the most challenging mathematical operation we encounter in ray tracing.

It is convenient to use the symbolic solver feature of Matlab to solve quadratic equations (see Problems 1.2–1.7). However, solution of Eq. (1.13) is relatively straightforward and provides an opportunity to point out certain useful properties of the quadratic coefficients. Upon carrying out the indicated operations and rearranging the result, we have

$$(L^2 + M^2 + N^2)\,t^2 + 2[L(x_0 - x_C) + M(y_0 - y_C) + N(z_0 - z_C)]\,t + (x_0 - x_C)^2 + (y_0 - y_C)^2 + (z_0 - z_C)^2 - R^2 = 0, \tag{1.14}$$

or

$$A\,t^2 + B\,t + C = 0, \tag{1.15}$$

where

$$A = L^2 + M^2 + N^2, \tag{1.16}$$

$$B = 2[L(x_0 - x_C) + M(y_0 - y_C) + N(z_0 - z_C)], \tag{1.17}$$

and

$$C = (x_0 - x_C)^2 + (y_0 - y_C)^2 + (z_0 - z_C)^2 - R^2. \tag{1.18}$$

The coefficients A, B, and C are defined in terms of known quantities and, thus, are themselves known. Equation (1.14) can now be solved for t, yielding

$$t_1 = (-B + \sqrt{B^2 - 4AC})/2A \text{ and } t_2 = (-B - \sqrt{B^2 - 4AC})/2A. \tag{1.19}$$

We define a **quadratic surface** as any surface whose algebraic equation $S(x, y, z) = 0$ is second-order. It turns out that, in addition to plane surfaces, essentially all enclosures of practical engineering interest include such surfaces or surfaces that can be subdivided into such surfaces. Listed in Table 1.1 are some of the quadratic surfaces commonly encountered in radiation heat transfer and applied optics. Equation (1.19), with generally different expressions for the coefficients A, B, and C, yields the candidate values of the distance t for all quadratic surfaces.

Note that Eq. (1.19) has two roots, t_1 and t_2. The physical interpretation of two roots is that a given ray "intersects" the spherical surface at two points. However, the intersection may be **degenerate** (both roots corresponding to the same point) or **imaginary** (the ray does not physically intersect the sphere). Only in the case of plane surfaces are single,

Table 1.1 Quadratic surfaces commonly encountered in radiation heat transfer and applied optics modeling.

Name	$S(x, y, z) = 0$	Notes
Sphere	$(x-x_C)^2+(y-y_C)^2+(z-z_C)^2-R^2=0$	R = radius, center at (x_C, y_C, z_C)
Tri-axial ellipsoid	$\frac{(x-x_C)^2}{a^2}+\frac{(y-y_C)^2}{b^2}+\frac{(z-z_C)^2}{c^2}-1=0$	a, b, c = semi-axes, center at (x_C, y_C, z_C)
Spheroid	$\frac{(x-x_C)^2}{a^2}+\frac{(y-y_C)^2}{a^2}+\frac{(z-z_C)^2}{c^2}-1=0$	Prolate if $c > a$, oblate if $c < a$, center at (x_C, y_C, z_C)
Elliptic paraboloid	$\frac{(x-x_C)^2}{a^2}+\frac{(y-y_C)^2}{b^2}-z=0$	Upward-opening, origin at (x_C, y_C, z_C)
Hyperbolic paraboloid ("Potato chip")	$(x/a)^2-(y/b)^2-z=0$	Opens up along x-axis, down along y-axis
Two-sheet hyperboloid	$(x/a)^2+(y/a)^2-(z/c)^2+1=0$	Rotationally symmetric about z-axis
Right-circular cone	$(x-x_v)^2+(y-y_v)^2 - \tan^2\alpha\,(z-z_v)^2=0$	Vertex at (x_v, y_v, z_v), α = cone half-angle
Right-circular cylinder	$x^2+y^2-R^2=0,\ 0\leq z\leq h$	Rotationally symmetric about z-axis, h = height

non-degenerate roots obtained. More than two roots arise in the case of higher-order surfaces but, as has already been pointed out, most enclosures of practical engineering interest are composed of either planes or quadratic surfaces.

The physical significance of quadratic roots is illustrated in Figure 1.3. If both roots are real ($B^2 > 4AC$) in Eq. (1.19), the ray emitted from P_0 intersects the surface at two points, P_1 and P_2, where one corresponds to the plus (+) sign in Eq. (1.19) and the other corresponds to the minus (−) sign. If $B^2 = 4AC$, the roots are degenerate and the single solution, $t = -B/2A$, is obtained. This signifies that the ray is tangent to the sphere at a single point P_3. Finally, if both roots are complex ($B^2 < 4AC$), the ray fails to intersect the surface.

At the most, only one real root is physically significant. How do we choose between the two available real roots? This is a trivial choice for someone provided with an image such as Figure 1.3; however, a computer requires an algorithm based on the values of the quadratic coefficients A, B, and C. In the special case of a sphere we recognize that $A = \sqrt{\boldsymbol{v} \cdot \boldsymbol{v}} = 1$, where the vector $\boldsymbol{v}$ is given by Eq. (1.3). While this simplifies the algebra somewhat, it does not otherwise contribute to the process of identifying the correct root. The coefficient B in the case of a sphere can be expressed

$$B/2 = \boldsymbol{v} \cdot \boldsymbol{V}_0, \tag{1.20}$$

where $\boldsymbol{V_0}$ is the vector directed from the center of the sphere (x_C, y_C, z_C) to the source point (x_0, y_0, z_0),

$$\boldsymbol{V}_0 = (x_0 - x_C)\,\boldsymbol{i} + (y_0 - y_C)\,\boldsymbol{j} + (z_0 - z_C)\,\boldsymbol{k}. \tag{1.21}$$

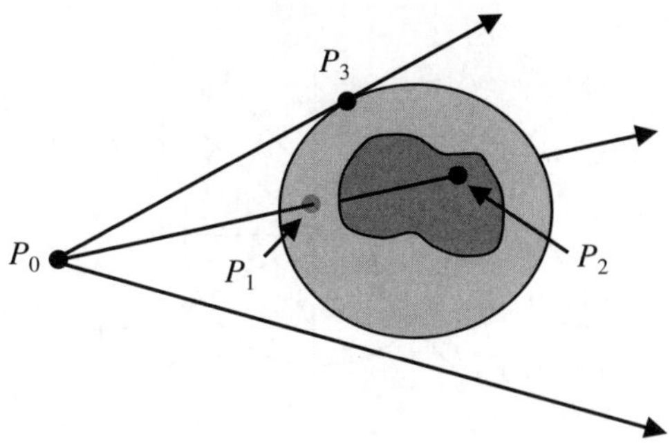

Figure 1.3 Possible disposition of three rays emitted from point P_0.

The value of the coefficient C determines whether the source point is interior to the sphere ($C<0$), exterior to the sphere ($C>0$), or lies in the surface of the sphere ($C = 0$). It has already been pointed out that the value of the argument of the square root in Eq. (1.19), $B^2 - 4AC$, determines whether or not the ray intersects the sphere. In the case of real roots, including degenerate roots, the interplay between the B and C coefficients determines the sign choice, as demonstrated in the following example.

Figure 1.4 depicts candidate rays in the interior of the hollow sphere defined by Eq. (1.12). The location of the origin of coordinates (0, 0, 0) is immaterial in the following argument. The ray source point (x_0, y_0, z_0), the sphere center (x_C, y_C, z_C) and radius R, and the components of the unit vector $\boldsymbol{v} = (L, M, N)$ are all assumed known. Because the ray source point is interior to the sphere, we find that $C<0$. It follows that $B^2 - 4AC>0$ regardless of the value of B. Therefore, there exist two real roots, corresponding to the two candidate intersections at (x_u, y_u, z_u) and (x_d, y_d, z_d).

The unit normal vector on the interior surface of the sphere is given by

$$\boldsymbol{n} = -\frac{\nabla S}{|\nabla S|} = -\left(\frac{x - x_c}{R}\right)\boldsymbol{i} - \left(\frac{y - y_c}{R}\right)\boldsymbol{j} - \left(\frac{z - z_c}{R}\right)\boldsymbol{k} = -\frac{1}{R}(\boldsymbol{V} + \boldsymbol{V}_0). \tag{1.22}$$

Then it can be demonstrated (see Problem 1.10) that

$$\boldsymbol{v} \cdot \boldsymbol{n} = -\frac{1}{R}\left(t + \frac{B}{2}\right). \tag{1.23}$$

In order for a ray emitted from point (x_0, y_0, z_0) in direction (L, M, N) to intercept the wall of the spherical shell at point (x, y, z), it must be true that $\boldsymbol{v} \cdot \boldsymbol{n} < 0$ (see Problem 1.11). We conclude that the correct sign in Eq. (1.19) is the one that leads to $t + b/2 > 0$; i.e., the "+" sign must be

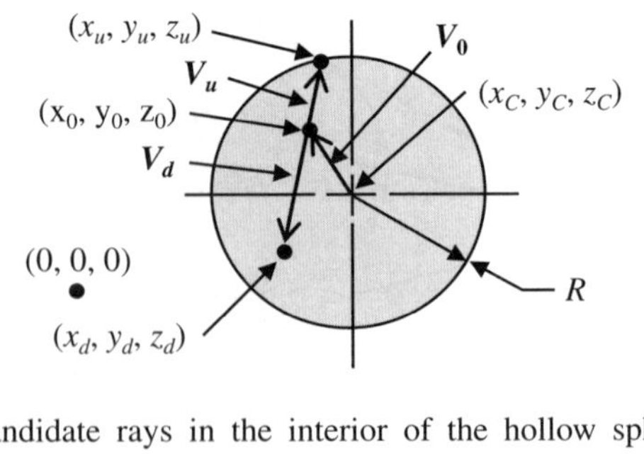

Figure 1.4 Candidate rays in the interior of the hollow sphere defined by Eq. (1.12).

selected in this case. It is emphasized that this is a general result for the case of a ray emitted from any position interior to a spherical enclosure.

The above example illustrates the important point that Eq. (1.5) describes an infinitely long line. In Figure 1.4 the upward-directed ray $\boldsymbol{V}_u$ and the downward-directed ray $\boldsymbol{V}_d$ both lie on the same line. Therefore, the simultaneous solution of the equations for this line with the equation for the sphere necessarily leads to two roots and, thus, to two intersections, only one of which is relevant. The true point of intersection is the **forward candidate**, i.e., the one lying in the direction of $\boldsymbol{v}$ with respect to the source point. The incorrect point, which lies in the opposite direction of $\boldsymbol{v}$, is referred to as a **back candidate**. Identification and elimination of back candidates is a constant concern in ray tracing.

Suppose the situation shown in Figure 1.4 is modified so that the source point (x_0, y_0, z_0) lies on the same line except that it now lies outside the sphere. In this case both points of intersection can be forward candidates or both can be back candidates, depending on the direction of the emitted ray. How do we sort out these possibilities?

In Figure 1.5a it is obvious to the human observer that both points of intersection are forward candidates, while in Figure 1.5b both are clearly back candidates. Furthermore, it seems equally obvious – once again to the human observer – that in the case of two forward candidates, the one nearer the source, and thus corresponding to the smaller value of t, is the correct choice. But, in fact, this would be true only if it is the outer surface of the sphere that is optically active. All surfaces, being the locus of all points where $S(x, y, z) = 0$, are vanishingly thin. Therefore, a given

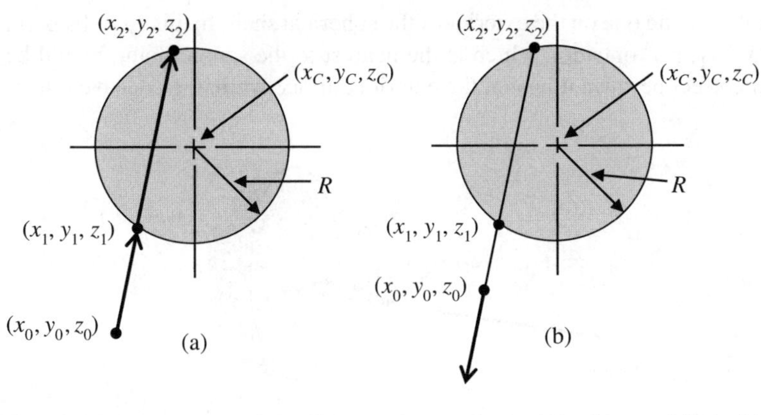

Figure 1.5 In (a) both points of intersection are forward candidates while in (b) both points of intersection are back candidates.

location (x_1, y_1, z_1) really represents *two* surfaces, or, more precisely, two sides of a single surface. Therefore, in Figure 1.5a the point (x_1, y_1, z_1) can represent either the inside surface or the outside surface of the sphere, depending on which surface is designated to be optically active. If the inside surface of the sphere in Figure 1.5a is deemed to be optically active, then the outside surface is mathematically transparent and point (x_2, y_2, z_2) rather than point (x_1, y_1, z_1) is the valid forward candidate.

The spherical sector in Figure 1.6 might represent a concave mirror used as an optical element, for example in a telescope. In this case its interior surface is considered to be a perfect mirror while its exterior surface, as well as the extended interior and exterior surfaces of the sphere from which the mirror is formed, indicated by the broken curve, simply do not exist. Given a ray originating at point (x_0, y_0, z_0), how do we determine (without looking at the figure) if it is reflected from the concave mirror?

Suppose simultaneous solution of Eqs. (1.5) and (1.12) yields the pair of real roots, t_1 and t_2, so that the corresponding coordinates for the candidate points are

$$x_1 = x_0 + Lt_1, y_1 = y_0 + Mt_1, \text{ and } z_1 = z_0 + Nt_1 \tag{1.24}$$

and

$$x_2 = x_0 + Lt_2, y_2 = y_0 + Mt_2, \text{ and } z_2 = z_0 + Nt_2. \tag{1.25}$$

How does the computer "know" which is the correct point? Bear in mind that these two intersections are actually two candidate points, one on the interior and one on the exterior of the spherical shell. In this case the point (x_1, y_1, z_1) (outside), which is the nearest to the source point, would be incorrect because it lies on the exterior surface while the concave mirror

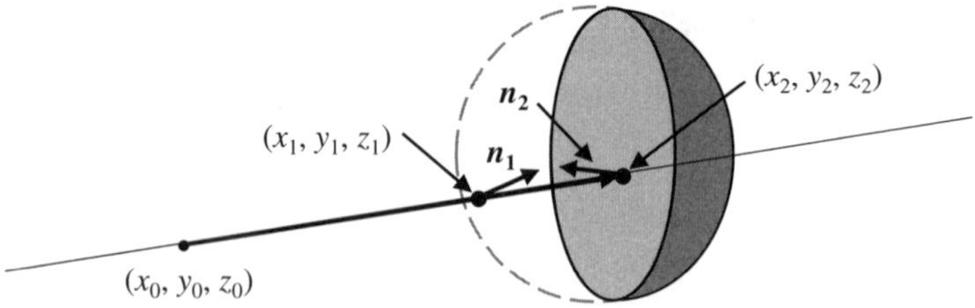

Figure 1.6 Instructing the computer to designate (x_2, y_2, z_2) rather than (x_1, y_1, z_1) as the point of intersection of the ray on the concave spherical sector.

is known to be a section of the interior surface. An essential feature of a successful Monte Carlo ray-trace (MCRT) program is that it must compute all intersections of all rays with all surfaces of the enclosure, and then sort through all of the resulting candidates to identify the right one. When the process arrives at the spherical-sector mirror shown in Figure 1.6, it initially treats it as a complete spherical shell having two exterior surfaces and two interior surfaces. Therefore, in the case of this particular surface, we must first instruct the computer to ignore the exterior surfaces and consider only the two interior surfaces.

How does the computer distinguish between exterior and interior surfaces? Firstly, it computes the local surface unit normal vector

$$\boldsymbol{n} = \pm\nabla S/|\nabla S|. \tag{1.26}$$

Following convention, the plus (+) sign is used in this calculation when the exterior surface is to be considered, while the minus (−) sign is used in the case of the interior surface. (The reader is cautioned that while the distinction between the exterior and the interior of a sphere is obvious, this is not always the case for many common quadratic surfaces.) In the example of Figure 1.6, the minus sign is used since the unit vector is pointed inward. Next the computer evaluates the sign of the dot product $\boldsymbol{v} \cdot \boldsymbol{n}$. We have already seen that the ray is directed away from any surface for which $\boldsymbol{v} \cdot \boldsymbol{n} > 0$ and is directed toward any surface for which $\boldsymbol{v} \cdot \boldsymbol{n} < 0$. Thus, in the current example, since $\boldsymbol{v} \cdot \boldsymbol{n}_2 < 0$, it is clear to both the computer and its human operator that the ray is incident to the interior point (x_2, y_2, z_2) rather than to the exterior point (x_1, y_1, z_1).

The foregoing example makes it clear that the development of ray-tracing skills requires familiarity with geometry, trigonometry, and algebra as well as an appreciation for logic and possession of computer programming skills. Indeed, a modestly gifted high-school senior could master the basic skill set in a matter of a few days. It is only the sophistication of the rules governing the interaction of electromagnetic radiation with materials of practical engineering interest that elevates the material in this book to the advanced university level.

1.4 Ideal Models for Emission, Reflection, and Absorption of Rays

To this point we have treated the ray as a strictly mathematical concept without considering its physical nature. However, as we move on to the phenomena of emission, absorption, reflection, scattering, and refraction, which occur when a ray intersects a surface, it will be convenient to introduce certain models borrowed from geometrical optics. In later chapters, we explore the principles of physical optics underlying these models. However, for the present it is convenient to exploit their relative simplicity as a tool for developing ray-tracing skills. This is not to say that the models introduced in this section are of pedagogical interest only; indeed, they have been the basis for traditional radiation heat transfer practice for the past century, during which time they have consistently yielded results whose accuracy is at least as good as that afforded by contemporary conduction and convection heat transfer epistemology.

We have been using the generally well understood term "surface" without formal definition. It is now appropriate to formally define a surface as the interface separating two regions of space having different optical properties. In fact, true surfaces do not exist, although approximations of surface behavior can be approached to an arbitrarily high degree of precision.

The optical behavior of a material substance is characterized by its **index of refraction** and its **extinction coefficient**. As a ray encounters the interface between two materials having different values of these optical properties, a portion of its power is redirected away from the interface. This portion of the incident power is said to be "reflected." Of the power that crosses into the second medium, a portion is said to be "absorbed" while the remainder is said to be either "scattered" or "refracted." The scattered and refracted power continues to propagate through the second medium while the absorbed power is locally converted into sensible heat. The two most prevalent models for describing reflection at a surface are the **specular reflection model** and the **diffuse reflection model**. These two models are important because they represent opposite extremes, both of which are often the desired behavior in engineering applications.

The specular reflection model, used to describe the ideal optical behavior of mirrors and certain other highly polished surfaces, is illustrated in Figure 1.7. In the figure, $\boldsymbol{v}$ and $\boldsymbol{n}$ are the unit vectors defined by Eqs. (1.3) and (1.26), respectively. The rule for specular reflection may be deduced by inspection of the figure; that is,

$$\boldsymbol{v}_R = \boldsymbol{v}_I - 2(\boldsymbol{v}_I \cdot \boldsymbol{n})\,\boldsymbol{n} = \boldsymbol{v}_I + 2\cos(\vartheta)\,\boldsymbol{n}, \qquad (1.27)$$

where $\vartheta = \vartheta_I = \vartheta_R$. Note that the specular reflection model says nothing about the power of the reflected ray compared to that of the incident ray, although in the case of an ideal mirror they are assumed to be the same.

The diffuse reflection model, commonly used to describe the optical behavior of all non-specular surfaces, is illustrated in Figure 1.8. In the

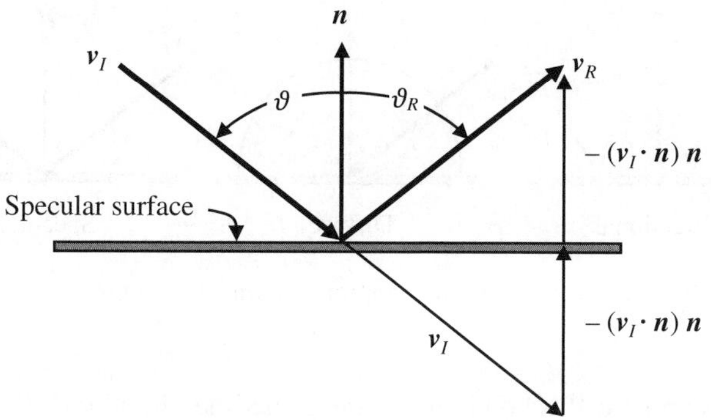

Figure 1.7 The specular reflection model.

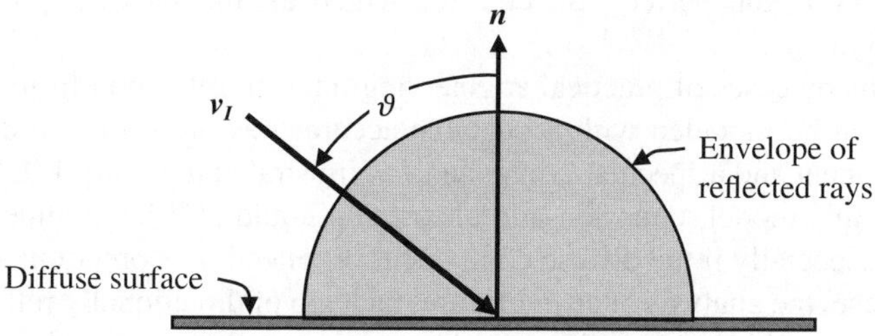

Figure 1.8 The diffuse reflection model.

MCRT interpretation of diffuse reflection, the reflected portion of the power contained in the incident ray v_I is redistributed into the hemispherical space above the point of incidence by a large but finite number of rays, where each ray carries the same power. In this case the directional distribution of the rays is governed by the values of a sequence of random numbers, as described later in this chapter. In principle, the number of directional components involved in such a reflection is infinite. However, in a practical ray trace, the number N of reflected components is necessarily finite, with each reflected component considered to carry a fraction $1/N$ of the total reflected power.

The diffuse reflection model may be extended to describe the directionality of emission and absorption. Theoretical considerations lead to the conclusion that if a surface is a diffuse reflector it must also be a diffuse absorber; that is the fraction of power absorbed from an incident ray is independent of the ray's direction. Furthermore, a diffuse absorber

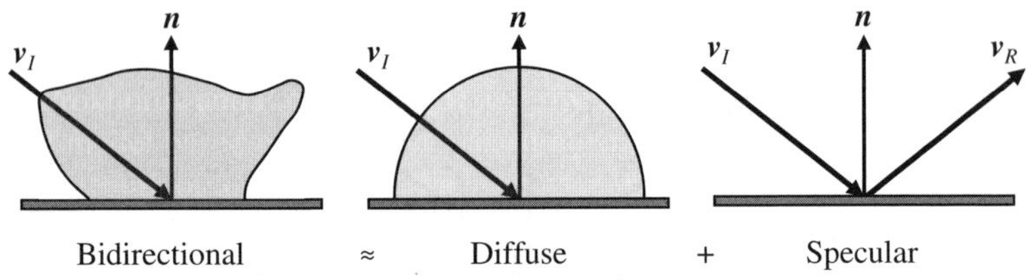

Figure 1.9 The diffuse–specular approximation of a bidirectional reflecting surface.

must also be a diffuse emitter, meaning that the directional distribution of emitted equal-power rays is governed by the same statistical rules that describe diffuse reflection. Finally, in the limiting case of a perfect absorber, all the incident radiation is absorbed. The opposite of a perfect absorber is the perfect reflector for which all the incident power is reflected.

In many cases of practical engineering interest, reflection from a surface can be modeled with acceptable accuracy as the sum of a diffuse component and a specular component, as illustrated in Figure 1.9. When using this model, emission and absorption could still be assumed diffuse, especially if the diffuse component of reflection is predominant. In any case, the analysis of an enclosure made up of directionally reflecting surfaces could always be improved by assuming diffuse–specular reflections, even if diffuse emission and absorption are retained.

More sophisticated models for emission, reflection, and absorption, which recognize that they are generally functions of both wavelength and direction, are considered in Chapter 4. The coefficients corresponding to these models provide sufficient information for rays to navigate from point to point along straight-line paths. The formal definition of the **bidirectional spectral reflectivity**, $\rho(\lambda, \vartheta_I, \varphi_I, \vartheta_R, \varphi_R)$, is given in Chapter 2. Its equivalent form in the MCRT environment addresses the question, "If a ray carrying one unit of power in wavelength interval $\Delta\lambda$ about wavelength λ is incident to a point on a surface from direction ϑ_I, φ_I, what fraction of the incident power is carried by a ray reflected from that point in direction ϑ_R, φ_R?" Its value for a specified combination of directions at a given wavelength is obtained by reference to a look-up table or to an empirical model. The bidirectional spectral reflectivity is mutually reciprocal in the sense that its incident and reflected directions can be interchanged without changing its value; i.e.,

$$\rho(\lambda, \vartheta_I, \varphi_I, \vartheta_R, \varphi_R) = \rho(\lambda, \vartheta_R, \varphi_R, \vartheta_I, \varphi_I). \quad (1.28)$$

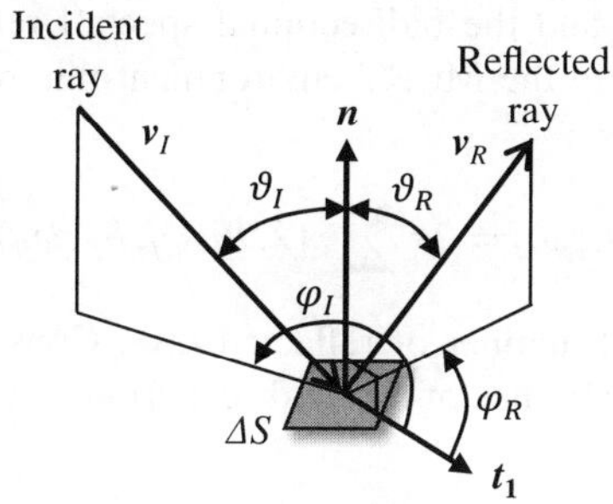

Figure 1.10 Nomenclature for incident and reflected rays.

We learn in Chapter 2 that all models for emission, reflection, and absorption follow from knowledge of the bidirectional spectral reflectivity.

Figure 1.10 illustrates the conventions for defining the incident and reflected directions. In the figure, the unit vector $\boldsymbol{t}_1$ is tangential to the surface element ΔS and is normal to the local unit normal vector $\boldsymbol{n}$.

It is often convenient to define the direction of the unit tangent vector such that

$$\boldsymbol{t}_1 = \boldsymbol{n} \times \boldsymbol{v}_I / |\boldsymbol{n} \times \boldsymbol{v}_I|, \tag{1.29}$$

since both $\boldsymbol{v}_I$ and $\boldsymbol{n}$ are known. However, the convention fails on those rare occasions when, to within the precision of the processor being used, $\boldsymbol{v}_I = -\boldsymbol{n}$. In this case, the denominator in Eq. (1.29) vanishes and the offending ray must either be discarded or slightly redirected. An alternative approach that avoids this weakness is presented in Chapter 3.

We define the **directional spectral emissivity** $\varepsilon(\lambda, \vartheta_E, \varphi_E)$ as the efficiency, compared to an ideal surface, with which a real surface emits thermal radiation, in wavelength interval $\Delta\lambda$ about wavelength λ, and in direction (ϑ_E, φ_E). The nature of the ideal surface referred to here is considered in Chapter 2. The **directional spectral absorptivity** $\alpha(\lambda, \vartheta_I, \varphi_I)$ is the fraction of incident thermal radiation in that wavelength interval, incident from direction (ϑ_I, φ_I), that is absorbed. It can be demonstrated that the directional spectral absorptivity is equal to the directional spectral emissivity at the same wavelength when $(\vartheta_I, \varphi_I) = (\vartheta_E, \varphi_E)$ [1].

Finally, the **directional-hemispherical spectral reflectivity** $\rho(\lambda, \vartheta_I, \varphi_I)$ is defined as the fraction of the power incident from direction (ϑ_I, φ_I) that is reflected into the hemispherical space above the point of incidence. The formal relationship between the directional-hemispherical spectral reflectivity and the bidirectional spectral reflectivity is developed in Chapter 2. In the MCRT environment this relationship can be expressed

$$\rho(\lambda, \vartheta_I, \varphi_I) = 2\pi \sum \rho(\lambda, \vartheta_I, \varphi_I, \vartheta_R, \varphi_R)/N, \tag{1.30}$$

where N is the total number of reflected rays. Conservation of energy requires that, in the absence of forward scattering and refraction (i.e., for an **opaque** surface),

$$\alpha(\lambda, \vartheta_I, \varphi_I) = 1 - \rho(\lambda, \vartheta_I, \varphi_I). \tag{1.31}$$

Implementation of the definitions for these coefficients in terms of models of varying degrees of sophistication is the topic of Chapter 4. For now, the relatively simple specular and diffuse reflection models illustrated in Figures 1.7 and 1.8 are well suited to the pedagogical purposes of the current chapter. According to these models, $\rho(\lambda, \vartheta_I, \varphi_I,) = \rho(\lambda$ *only)*, with $\vartheta_R = \vartheta_I$ and $\varphi_R = \varphi_I + \pi$ for specular surfaces and, for diffuse surfaces,

$$\vartheta_R = \sin^{-1}[\sqrt{R_\vartheta}] \text{ and } \varphi_R = 2\pi R_\varphi, \tag{1.32}$$

where in both cases the angles ϑ and φ are measured with respect to normal and tangent unit vectors, as indicated in Figure 1.10. In Eq. (1.32), R_ϑ and R_φ are random numbers uniformly distributed between zero and unity. Random number generation is treated more fully in the Appendix but the random-number generators available in most computing environments, for example *rand* in Matlab, are adequate for most applications.

Models for the remaining coefficients follow from their definitions. Thus, for opaque surfaces the directional spectral absorptivity $\alpha(\lambda, \vartheta_I, \varphi_I)$ becomes $\alpha(\lambda) = 1 - \rho(\lambda)$, and the directional spectral emissivity $\varepsilon(\lambda, \vartheta_E, \varphi_E)$ becomes $\varepsilon(\lambda) = \alpha(\lambda)$. In the case of a diffuse source the direction of emission is determined using Eq. (1.32) while, in the case of a directional source, rays are emitted with some specified directional distribution. For example, in the special case of a **collimated** source, *all* rays are emitted in a specified direction (ϑ_E, φ_E).

Bearing in mind that rays navigate using direction cosines, it would be useful if Eq. (1.32), with the "*R*" subscripts replaced with "*E*" subscripts, could be used to compute L, M, and N for the case of diffuse emission. For this purpose a second unit tangent vector $\boldsymbol{t}_2$ that is normal to both $\boldsymbol{n}$ and $\boldsymbol{t}_1$, is defined

$$\boldsymbol{t}_2 = \boldsymbol{n} \times \boldsymbol{t}_1. \tag{1.33}$$

It can then be demonstrated (see Problem 1.12) that

$$L = n_x \cos\vartheta_E + t_{1,x} \sin\vartheta_E \cos\varphi_E + t_{2,x} \sin\vartheta_E \sin\varphi_E,$$

$$M = n_y \cos\vartheta_E + t_{1,y} \sin\vartheta_E \cos\varphi_E + t_{2,y} \sin\vartheta_E \sin\varphi_E,$$

$$\text{and } N = n_z \cos\vartheta_E + t_{1,z} \sin\vartheta_E \cos\varphi_E + t_{2,z} \sin\vartheta_E \sin\varphi_E. \quad (1.34)$$

Finally, in some situations it is justified to assume that the surface model coefficients ε, α, and ρ are independent of wavelength. This so-called **graybody approximation** has dominated radiation heat transfer practice for the past century. Its enduring popularity stems from the fact that it significantly simplifies the analysis while remaining relevant in a wide range of practical applications. Therefore, in this introductory chapter to ray tracing we temporarily set aside the analytical complications imposed by wavelength-dependent surface models with the idea of adding them later as needed.

1.5 Scattering and Refraction

The power incident to a surface that is neither absorbed locally nor reflected is then either scattered in a process analogous to reflection, or it is refracted. In the MCRT description of radiation heat transfer, scattering is modeled by subdividing the incident ray into many equal-power rays, with each scattered ray continuing in a direction determined by an appropriate scattering model. The complex phenomenon of scattering is treated in detail in Chapter 5, which deals with radiation propagating through a **participating medium**. The simplest and most basic model for scattering, which is used in the early chapters of this book, is the assumption that scattering can be neglected, as is often the case in radiation heat transfer.

In the ray-trace description of geometrical optics, refraction refers to the abrupt change in direction of the transmitted ray as it passes through an interface. The Snell–Descartes law, illustrated in Figure 1.11, represents reality very well, especially for interfaces between air and common materials used in the fabrication of lenses, filters, retarder plates, and windows. According to the Snell–Descartes law

$$\sin(\vartheta_1)/\sin(\vartheta_2) = n_2/n_1, \quad (1.35)$$

where n_1 and n_2 are the refractive indices of the two materials whose interface provokes refraction. Problem 1.14 is an important application of this principle in applied optics.

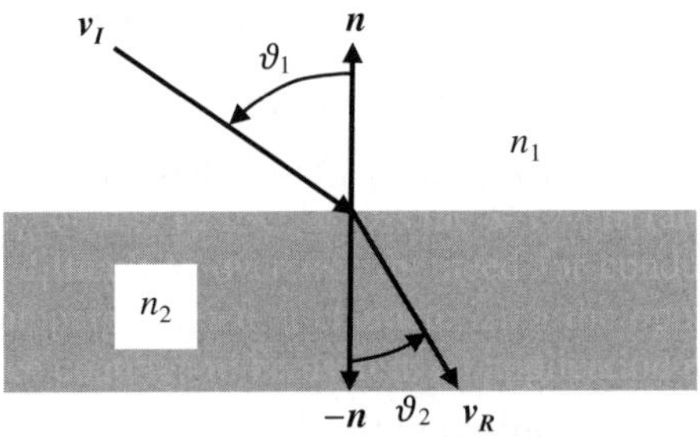

Figure 1.11 Illustration of the Snell–Descartes law of refraction.

1.6 Meshing and Indexing

The MCRT method used throughout this book requires that the modeling space be subdivided into surface and volume elements, i.e., that it be appropriately **meshed**. While entire books have been written on this topic alone, the limited treatment offered here is adequate for the needs of most MCRT analyses. The meshes used in the MCRT method must be amenable to **indexing**. Indexing refers to the process of systematically numbering the surface and volume elements in such a way that the numbers, called **indices**, can be determined algorithmically from the coordinates of a point lying on a surface element or within a volume element.

Pedagogical considerations favor limitation of the discussion presented here to rectilinear spaces, i.e., to spaces that can be represented by rectangular solid blocks. As used here, the word "solid" implies only that the spaces are three-dimensional. Many, if not most, enclosures of practical engineering interest can be accurately represented using a rectilinear mesh if care is taken to ensure that the surface element unit normal vectors represent the actual local curvature. The methods presented in this section can be extended to spaces consisting of trapezoidal, cylindrical (both circular and noncircular), and spheroidal solids.

Consider the hollow, three-dimensional rectilinear space illustrated in Figure 1.12. Use of the MCRT method often requires that the space be divided into N surface elements and $n - N$ volume elements, with a unique number, or "index", algorithmically assigned to each element. Furthermore, square surface elements and cubic volume elements are highly desirable. Finally, the resulting mesh must be sufficiently dense to assure adequate spatial resolution of the results obtained using an MCRT analysis. How do we go about satisfying all of these requirements? Consider the following numerical examples.

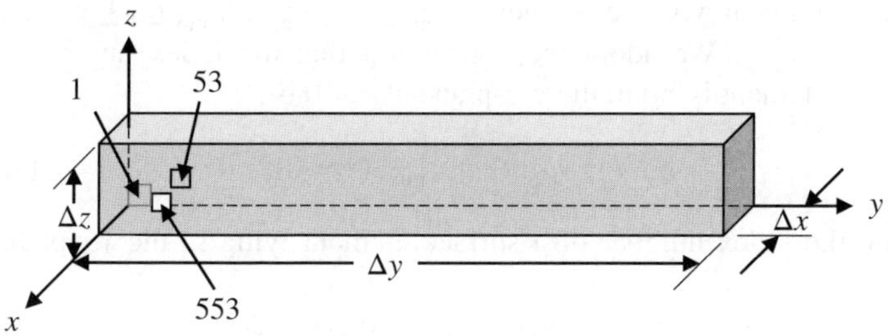

Figure 1.12 A hollow, three-dimensional rectilinear space.

Example 1.1

Suppose in Figure 1.12 that $\Delta x = 5.00$, $\Delta y = 50.00$, and $\Delta z = 10.00$, where the length units are arbitrary. The corresponding meshing solution, referred to as the **natural** solution, has five divisions in the x-direction ($n_x = 5$), 50 divisions in the y-direction ($n_y = 50$), and 10 divisions in the z-direction ($n_z = 10$). This meshing produces $2(n_x n_y + n_x n_z + n_y n_z) = N = 1600$ perfectly square surface elements and $n_x n_y n_z = n - N = 2500$ perfectly cubic volume elements. If it is found that the natural solution is too coarse to provide adequate spatial resolution in an eventual MCRT analysis, we simply multiply the number of divisions in each direction by the same integer to produce an acceptably dense mesh. On the other hand, if the mesh density is greater than needed – and therefore inefficient— we can always reduce it by dividing the number of divisions in each direction by n_x, n_y, or n_z, depending on which of the three values is smallest. This produces the minimal mesh that preserves the desirable property of square surface elements aligned with cubic volume elements. Such minimal meshes are referred to as **primitives**.

The general requirements for indexing are:

(i) element indices should be assigned algorithmically based on the coordinates of one of the element corners and, once assigned,
(ii) they should be algorithmically recoverable from the coordinates of any point on or within the element. The algorithm used for meshing is usually identical to the one used for indexing, but is used differently.

How do we index the space illustrated in Figure 1.12? All of the volume elements have six bounding surfaces, any one of which may be either real (i.e., a surface element) or imaginary (i.e., an interface between two adjoining volume elements). We have for any point (x, y, z) lying within a given volume element; $x_{min} \leq x \leq x_{max}$, $y_{min} \leq y \leq y_{max}$, and $z_{min} \leq z \leq z_{max}$. We adopt the convention that the index number of a surface element lying in the y, z-plane at $x = 0$ is

$$i = y_{max} + n_y\, z_{min}, (x = 0); \tag{1.36}$$

while the index number of a surface element lying in the y,z-plane at $x = \Delta x$ is

$$i = y_{max} + n_y\, z_{min} + n_y n_z, (x = \Delta x). \tag{1.37}$$

All of the surface elements are assumed to face the interior of the rectangular solid. Note that the corner values of x, y, and z are all integers, thus assuring that the indices themselves are integers. Then, in the current example ($n_y = 50$ and $n_z = 10$),

$$\begin{aligned} i &= 1 \text{ when } y_{max} = 1.00, z_{min} = 0.00, \text{ and } x = 0.00; \\ i &= 53 \text{ when } y_{max} = 3.00, z_{min} = 1.00, \text{ and } x = 0.00; \\ \text{and} \quad i &= 553 \text{ when } y_{max} = 3.00, z_{min} = 1.00, \text{ and } x = 5.00. \end{aligned}$$

Surface elements 1, 53, and 553 are indicated in Figure 1.12. It is left to the reader to puzzle out the expressions for the remaining surface element index numbers (Problem 1.20).

The corresponding expression for indexing the volume elements is

$$\begin{aligned} i &= N + x_{max} + n_x\,(y_{max} - 1) + n_x\, n_y\,(z_{max} - 1) \\ &= 1600 + x_{max} + 5\,(y_{max} - 1) + 250\,(z_{max} - 1). \end{aligned} \tag{1.38}$$

Derivation and validation of this result are left as an exercise (Problem 1.21). The process is further exemplified in Step 7, Section 5.3.

The MCRT method also requires that, given the x,y,z-coordinates somewhere on the walls of an enclosure, we be able to algorithmically recover the index number of the corresponding surface element. The logic that accomplishes this in the current example for surfaces in the y,z-plane at $x = 0$ is (see Problem 1.22):

$$\begin{aligned} &\text{for } x = 0; \text{if } y \neq 0 \text{ then } i = 1 + \mathit{floor}(y) + n_y\,\mathit{floor}(z), \\ &\qquad \text{else if } y = 0 \text{ then } i = 1 + n_y\,\mathit{floor}(z), \end{aligned} \tag{1.39}$$

where the *floor* operator in Matlab returns an integer equal to the real value of the argument truncated at the decimal point. For example, *floor*(5.57) = 5 and *floor*(0.57) = 0. Derivation and validation of the expressions for indexing the remaining five walls and for indexing the volume element containing the point (x, y, z) are left as exercises (Problem 1.23).

Example 1.2

While the foregoing is an excellent "textbook" example, it is not very realistic. It is unlikely that the natural meshing solution will be exact. A more general and far more realistic case might be $\Delta x = 5.50$, $\Delta y = 51.25$, and $\Delta z = 9.75$. Note that, as in the foregoing example, the dimensions are given to one-hundredth of a unit. We begin by scaling the dimensions with respect to the smallest dimension, $\Delta x = 5.50$ in this case. Thus $\Delta x' = \Delta x/\Delta x = 1.00$, $\Delta y' = \Delta y/\Delta x = 51.25/5.50 = 9.32$, and $\Delta z' = \Delta z/\Delta x = 9.75/5.50 = 1.77$, where each result has been rounded to the nearest one-hundredth of a unit. Then the "natural" meshing in this case is $n_x = 100$, $n_y = 932$, and $n_z = 177$. As in the previous example, this meshing produces square surface elements and cubic volume elements. Furthermore, the elements are one unit on a side to within the precision of the given dimensions. Note that the natural meshing turns out to be the same as the primitive meshing in this example. However, if n_z had turned out to be 178 rather than 177, the primitive mesh would have been $n_x = 50$, $n_y = 466$, and $n_z = 89$. Finally, indexing proceeds in this general case just as in the special case considered in Example 1.1, since x_{min}, y_{min}, z_{min}, x_{max}, y_{max}, and z_{max} are all integers.

We are now mathematically prepared to formulate and solve realistic radiation heat transfer and optics problems using the MCRT method. However, before moving directly into the main thrust of the book, we need to first establish fundamental knowledge of thermal radiation.

Problems

1.1 A plane whose unit normal vector $\boldsymbol{n} = \boldsymbol{k}$ intersects a sphere of radius R whose center is at (0, 0, 0). The radius of the disk formed by this intersection is $R/2$.

(a) In terms of R, what is the z-coordinate of the center of the disk?

(b) Determine if a ray emitted from $P_0 = (-R/2, 0, 0)$ in direction $(L, M, N) = (\sqrt{2}/2, 0, \sqrt{2}/2)$ intersects the disk. If so, where?

(c) In terms of R, where does the ray intersect the "cap" of the sphere lying above the disk?

1.2 A prolate spheroid (see Table 1.1) has a minor semi-axis of a, a major semi-axis of c (with $c > a$), and its center at $(0, y_C, 0)$. An x,z-plane intersects the spheroid at $0 < y_1 < a$, and a second x,z-plane intersects the spheroid at $y_1 < y_2 < a$. The intersections between the planes and the spheroid are ellipses with respective minor semi-axes of x_1 and x_2, and major semi-axes of z_1 and z_2. The values of x_1, y_1, z_1, x_2, y_2, and z_2 are considered to be known, and the values of a, b, and y_C are considered unknown. Derive explicit expressions for each of the three unknown quantities. Hint: You may want to make use of the Matlab symbolic solver feature.

1.3 Use the Matlab symbolic solver feature to solve Eq. (1.13) for t in terms of x_0, y_0, z_0, L, M, N, and R. Isolate and identify the coefficients A, B, and C in the solution.

1.4 Repeat Problem 1.3 for the case of a tri-axial ellipsoid whose center is located at (x_C, y_C, z_C). Use the symbols given in Table 1.1.

1.5 Repeat Problem 1.3 for the case of a two-sheet hyperboloid whose center is located at (x_C, y_C, z_C). The distance between the two vertices is $2c$. Use the symbols given in Table 1.1.

1.6 Repeat Problem 1.3 for the case of a right-circular cone whose center is located at (x_V, y_V, z_V). Use the symbols given in Table 1.1.

1.7 Repeat Problem 1.3 for the case of a right-circular cylinder whose center is located at (x_C, y_C, z_C). Use the symbols given in Table 1.1.

1.8 Demonstrate that if the quadratic coefficient $C = 0$ in Eq. (1.18), the line segment from point (x_0, y_0, z_0) to point (x_1, y_1, z_1) is a chord of the sphere.

1.9 The equation of a right-circular cone whose axis is parallel to the z-axis is given in Table 1.1. Consider the special case where $(x_v, y_v) = (0, 0)$, $z_v = 10$, and $\alpha = \pi/4$. A ray is emitted from $(x_0, y_0, z_0) = (-4, 2.5, 0)$ with direction cosines $(L, M, N) = (\sqrt{0.19}, 0, 0.9)$.

(a) Derive the equations for the quadratic coefficients A, B, and C.
(b) Solve the resulting quadratic equation for t_1 and t_2.
(c) Does the ray intersect the cone? If so, determine the coordinates of the candidate point or points of intersection.
(d) If two real points of intersection are found and only the inside surface of the cone is optically active, what are the direction

cosines of the reflected ray if the inside surface is an ideal mirror?

(e) If two real points of intersection are found and only the outside surface of the cone is optically active, what are the direction cosines of the reflected ray if the outside surface is an ideal mirror?

(f) For Parts (d) and (e), sketch the relevant cone surfaces and the incident and reflected rays.

1.10 Use Eqs. (1.20) and (1.22) to derive Eq. (1.23).

1.11 Verify that in order for a ray emitted from point (x_0, y_0, z_0) in direction (L, M, N) to intercept the wall of the spherical shell at point (x, y, z) it must be true that $\boldsymbol{v} \cdot \boldsymbol{n} < 0$.

1.12 Derive the three expressions in Eq. (1.34).

1.13 A hemispherical interior surface of radius 10 units is entirely located in the upper half-space ($z \geq 0$) with its center located at the origin of coordinates, (0, 0, 0), as shown in Figure 1.13. Write a computer program using Matlab (or equivalent) that traces 10 000 diffusely distributed rays from the origin into the hemisphere.

(a) Plot the points of intersection on the hemispherical surface.

(b) Show that the hemispherical surface can be subdivided into 10 equal-area bands by slicing it with nine parallel x,y-cutting planes at $z = 1, 2, 3, \ldots, 9$ units.

(c) Count the number of ray intersections in each band and then create a bar graph of these numbers as a function of the angle ϑ of the center of each band.

1.14 A condensing lens can be formed from the intersection of two spherical sectors of optical glass, as illustrated in Figure 1.14.

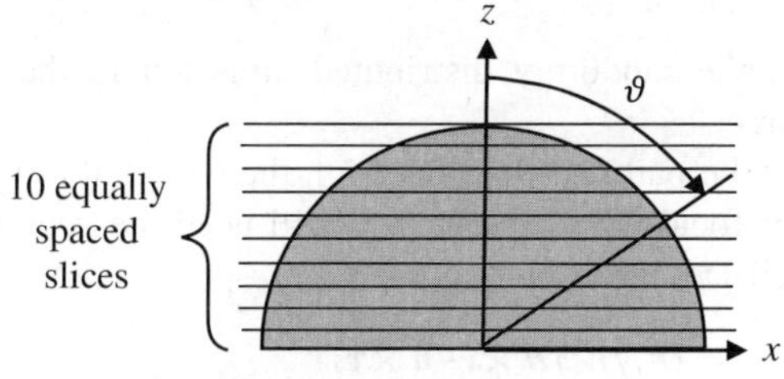

Figure 1.13 Geometry for Problem 1.13.

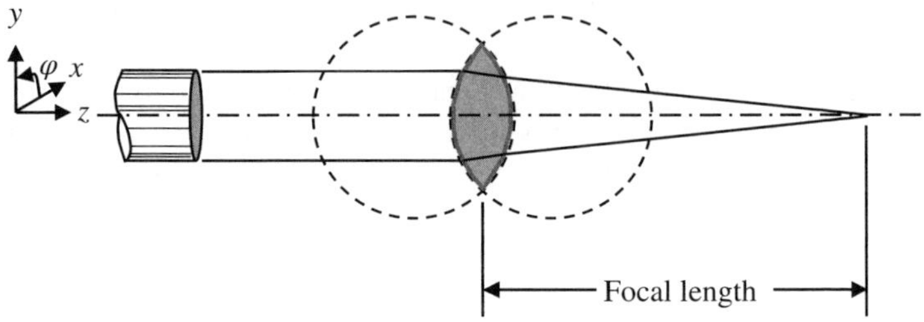

Figure 1.14 A condensing lens formed from two intersecting spherical sectors of optical glass.

The refractive index of air is 1.002772 and the refractive index of the optical glass used to construct the lens is 1.51681. The radius of curvature of both spherical sectors is 76.6 mm and the lens thickness at its center is 4.10 mm. A 5.0-mm diameter collimated beam is incident to the lens. A spatially uniform beam may be well simulated by emitting 1000 rays parallel to the optical axis of the lens at random x,y-locations in the source plane. If the center of the circular source of diameter D is located at $(x, y, z) = (0, 0, 0)$, the radial distance from the center is

$$r = (D/2)\ \sqrt{R_r}, \tag{1.40}$$

where R_r is a random number uniformly distributed between 0 and 1. A random angle φ, measured with respect to the x-axis, may be defined by a second random number, R_φ,

$$\varphi = 2\pi R_\varphi. \tag{1.41}$$

Then the x,y-coordinates of a random emission site are

$$x = r\ \cos\varphi, y = r\ \sin\varphi, \text{and } z = 0. \tag{1.42}$$

(a) Plot the randomly distributed emission in the plane of the source.

(b) Plot the paths of the rays from the source through the lens to their focal point. [Hint: You will need the vector form of the Snell-Descartes law,

$$\begin{aligned} v_{\boldsymbol{R}} &= (n_1/n_2)\,\boldsymbol{n} \times (-\boldsymbol{n} \times \boldsymbol{v_I}) \\ &\quad - \boldsymbol{n}\,[1 - (n_1/n_2)^2\,(\boldsymbol{n} \times \boldsymbol{v_I}) \cdot (\boldsymbol{n} \times \boldsymbol{v_I})]^{1/2}, \end{aligned} \tag{1.43}$$

where $\boldsymbol{n}$ is the local surface unit normal vector, $\boldsymbol{v}_I$ is the incident vector, and $\boldsymbol{v}_R$ is the refracted vector.]

(c) What is the focal length of this lens?

1.15 Derive the vector form of the Snell–Descartes law, Eq. (1.43).

1.16 The equation of a certain right-circular cone is

$$S(x, y, z) = x^2 + y^2 - (z - z_v)^2 \tan^2\alpha = 0, \tag{1.44}$$

where z_v is the z-location of its vertex and α is its half-angle.

(a) Use the Microsoft Word "draw" feature (Insert→Shapes) to sketch the full cone, showing both branches, or "sheets." Label the x-, y-, and z-axes and the angle α.

(b) Derive the expression for the inward-directed unit normal vector $\boldsymbol{n}$ for the upper sheet. By convention, the inward-directed unit normal vector for the upper sheet has a positive z-component.

(c) Derive the expression for the unit tangent vector $\boldsymbol{t}_1$ for the upper sheet that lies entirely in the surface and points toward the vertex.

(d) Derive the expression for the second unit tangent vector $\boldsymbol{t}_2$ that is mutually orthogonal to both the unit normal vector and the first unit tangent vector. Assure that the three vectors define a right-hand system.

(e) Add the three unit vectors to your sketch.

1.17 The equation of a right circular cylindrical surface is

$$S(x, y) = x^2 + y^2 - R^2 = 0, \tag{1.45}$$

where R is the radius.

(a) Use the Microsoft Word "draw" feature (Insert→Shapes) to sketch a right circular surface between $-L \leq z \leq L$. Label the x-, y-, and z-axes and the radius R.

(b) Derive the expression for the inward-directed unit normal vector $\boldsymbol{n}$.

(c) Derive the expression for the unit tangent vector $\boldsymbol{t}_1$ that lies entirely in the surface and points in the positive-z direction.

(d) Derive the expression for the second unit tangent vector $\boldsymbol{t}_2$ that is mutually orthogonal to both the unit normal vector and the first unit tangent vector. Assure that the three vectors define a right-hand system.

(e) Add the three unit vectors to your sketch.

1.18 The equation for an elliptic paraboloid is

$$S(x, y, z) = x^2 + y^2 - a(z - z_v) = 0, \tag{1.46}$$

where $a(z - z_v)$ is the radius of its "rim" at a distance z from its vertex at z_v.

(a) Use the Microsoft Word "draw" feature (Insert→Shapes) to sketch an elliptic paraboloid. Label the x-, y-, and z-axes.

(b) Derive the expression for the inward-directed unit normal vector $\boldsymbol{n}$.

(c) Derive the expression for the unit tangent vector $\boldsymbol{t_1}$ that lies entirely in the x,y-plane and points in the counter-clockwise direction looking down on the x,y-plane.

(d) Derive the expression for the second unit tangent vector $\boldsymbol{t_2}$ that is mutually orthogonal to both the unit normal vector and the first unit tangent vector. Assure that the three vectors define a right-hand system.

(e) Add the three unit vectors to your sketch.

1.19 Consider the wedge-shaped cavity depicted in Figure 1.15.

(a) Show that the equation for the sloped surface is

$$S(y, x) = z - (y_{\text{max}} - y)/\tan\alpha = 0. \tag{1.47}$$

(b) Derive the expression for the inward-directed unit normal vector $\boldsymbol{n}$ on the sloped surface.

(c) Derive the expression for the first unit tangent vector $\boldsymbol{t_1}$ on the sloped surface.

(d) Suppose $\alpha = 40°$ and that, during a Monte Carlo ray-trace, angles of $\vartheta = 34.6°$ and $\varphi = 102.4°$ are obtained for the reflection of a given ray from the sloped surface. What are the direction cosines L, M, N of the path of the reflected energy bundle?

(e) Suppose the point of reflection corresponding to Part (d) lies at the center of the sloped surface and that $y_{\text{max}} = 180.5\ \mu\text{m}$. Which surface does the reflected energy bundle strike and what are the coordinates of the point of intersection of the ray with the surface? Show the ray on a reasonably scaled drawing.

1.20 Consider the hollow, three-dimensional rectilinear space of Figure 1.12. Derive the expressions for the remaining surface element index numbers i following Eqs. (1.36) and (1.37).

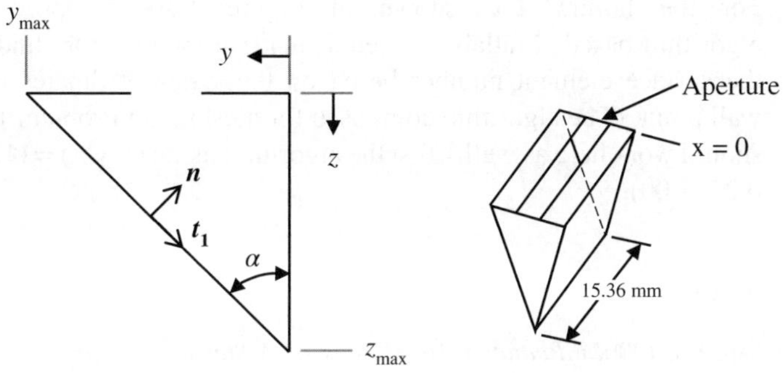

Figure 1.15 The wedge-shaped cavity of Problem 1.19.

1.21 Derive Eq. (1.38).

1.22 Derive Eq. (1.39).

1.23 Derive and validate the expressions for indexing the remaining five walls, as mentioned below Eq. (1.39).

1.24 Derive the expression for computing the index of the volume element containing the point (x, y, z) following Eq. (1.37).

1.25 Figure 1.16 shows a hollow box whose walls have been subdivided into four square surface elements. Compose and test an algorithm-based Matlab (or equivalent) function to index the surface elements of the box based on knowledge of the appropriate corner coordinates when $\Delta x = \Delta y = \Delta z = 2.00$. Use the MS Word drawing tool to sketch and fully label the box for both the natural and the primitive meshing schemes.

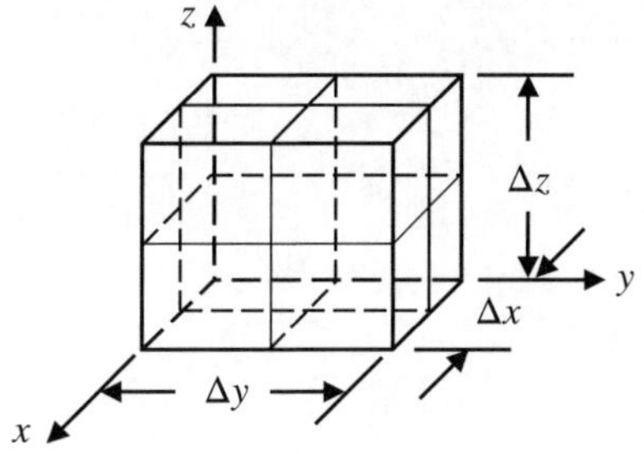

Figure 1.16 Geometry for Problems 1.25 and 1.26.

1.26 For the hollow box shown in Figure 1.16, compose an algorithm-based Matlab (or equivalent) function for finding the surface element number based on the x,y,z-coordinates of a wall point. (The algorithm composed for meshing in Problem 1.25 should work here as well.) Test the algorithm using $(x, y, z) = (1.45, 0.27, 2,00)$.

Reference

1. Mahan, J.R. (2002). *Radiation Heat Transfer: A Statistical Approach*. New York: Wiley.

2

Fundamentals of Thermal Radiation

Thermal radiation is emitted from any body whose temperature exceeds absolute zero. It is the mechanism by which heat is exchanged among bodies at different temperatures in the absence of an intervening medium. In this chapter we present the basic concepts from electromagnetic theory and geometrical optics that are required for the formulation of mathematical models describing radiation heat transfer. Included is an introductory treatment of the radiation behavior of dielectrics and metals.

2.1 Thermal Radiation

Electromagnetic (EM) waves, whose properties are explored in this chapter, carry energy from one location to another, even – indeed, especially – in a vacuum. The mathematical form of the magnitude of the electric field component of an EM wave propagating along the x-axis is

$$E(x,t) = E_0 e^{i(kx-\omega t)}, \tag{2.1}$$

where E (V m^{-1}) is the instantaneous electric field strength at position x (m) and time t (s), E_0 is the amplitude of its oscillation, k (m^{-1}) is

The Monte Carlo Ray-Trace Method in Radiation Heat Transfer and Applied Optics, First Edition. J. Robert Mahan.

This Work is a co-publication between John Wiley & Sons Ltd and ASME Press.
Companion website: www.wiley.com/go/mahan/monte-carlo

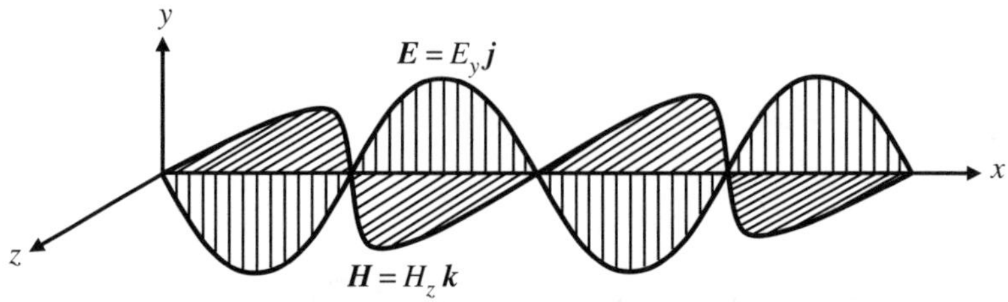

Figure 2.1 A y-polarized EM wave propagating along the x-axis.

the wave number, and ω (r s^{-1}) is the angular frequency. An analogous equation can be written for the magnetic field component, H (A m^{-1}). Figure 2.1 illustrates a y-polarized EM wave propagating along the x-axis. The power carried by this EM wave is given by the Poynting vector,

$$\boldsymbol{P} = \boldsymbol{E} \times \boldsymbol{H} = E_y \boldsymbol{j} \times H_z \boldsymbol{k} = E_y H_z \boldsymbol{i} \,(\mathrm{W/m^2}). \tag{2.2}$$

This is the basic mechanism of radiation heat transfer.

The frequency $\upsilon = \omega/2\pi$ (s^{-1}) of an electromagnetic wave is determined at its origin and does not change as it propagates. However, its wavelength $\lambda = 2\pi/k$ (m) varies according to the speed of light c (m s^{-1}) in the medium through which it propagates according to

$$\lambda = c/\upsilon = c_0/n\upsilon, \tag{2.3}$$

where $n \equiv c_0/c$ is the **index of refraction** and c_0 is the speed of light in a vacuum (~2.9979 × 10^8 m s^{-1}).

Figure 2.2 represents an idealized view of the electromagnetic spectrum for light propagating in a vacuum ($n = 1$). **Thermal radiation**, which includes visible radiation, is the part of the EM spectrum normally associated with radiation heat transfer. The spectral limits shown here

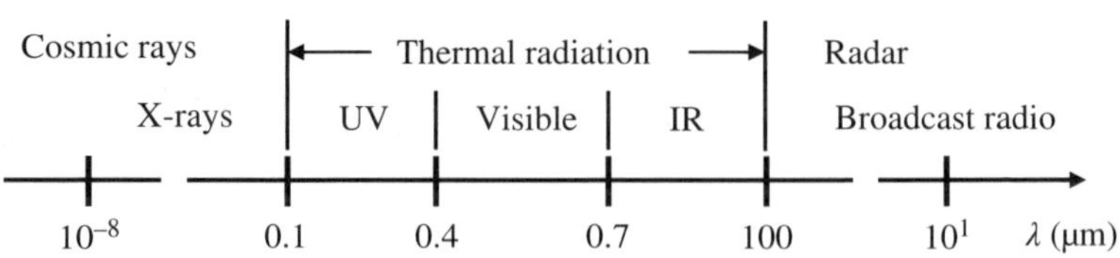

Figure 2.2 The electromagnetic spectrum.

for thermal radiation, while admittedly somewhat arbitrary, include more than 99% of the total energy emitted by the Sun.

2.2 Terminology

Further discussion of the nature of thermal radiation requires careful definition of certain concepts and terms of art. Chief among these is the concept of the **solid angle** Ω (sr), whose thorough understanding is critical to the study of both radiation heat transfer and applied optics. Consider Figure 2.3, which shows a differential surface element dS located a distance r from a second differential area dA. The line of length r connecting dA and dS intersects the normal to dA at an angle β_A, and it intersects the normal to dS at an angle β_S. As a concession to clarity, the surface elements dA and dS are necessarily drawn as finite in size but, in fact, both are arbitrarily small compared to the finite distance r. The surface element $dS \cos \beta_S$, which is hinged to surface element dS along their common lower edge, is tilted toward dA, so that the line r is normal to $dS \cos \beta_S$. Because both dS and $dS \cos \beta_S$ are vanishingly small, the distance between the points where r intersects them is negligible compared to the finite length r. Then the differential solid angle $d\Omega_S$ subtended by $dS \cos \beta_S$ at dA is defined

$$d\Omega_S \equiv \frac{dS \cos \beta_S}{r^2}. \tag{2.4}$$

Note that the solid angle in steradians (sr) is actually a dimensionless ratio of area over length squared, just as a one-dimensional angle in radians (r) is a dimensionless ratio of lengths.

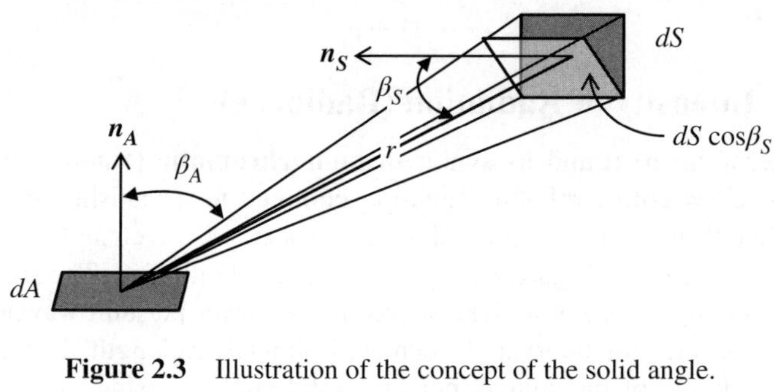

Figure 2.3 Illustration of the concept of the solid angle.

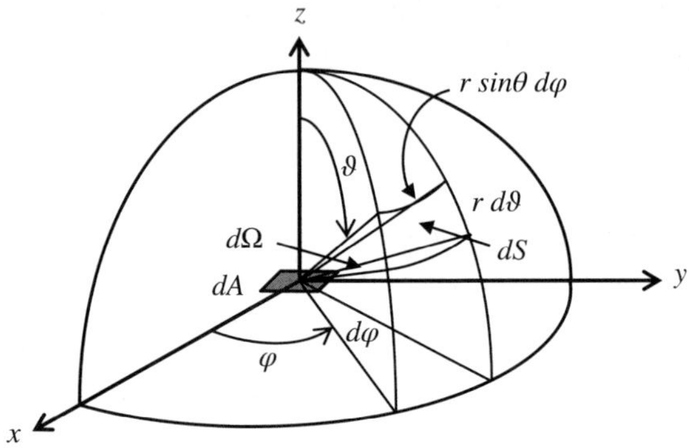

Figure 2.4 Differential solid angle subtended by a differential spherical sector at the center of a sphere.

An important special case of the solid angle is illustrated in Figure 2.4, which shows the differential solid angle subtended by a differential spherical sector dS at the center of a sphere of radius r. The projected area of the differential spherical sector when viewed from the center of the sphere is

$$dS \cos \beta_S = (r \sin \vartheta \, d\varphi)(r \, d\vartheta) = r^2 \sin \vartheta \, d\vartheta \, d\varphi \tag{2.5}$$

since $\beta_S = 0$. Then from Eq. (2.4) the differential solid angle in terms of ϑ and φ is

$$d\Omega = \sin \vartheta \, d\vartheta \, d\varphi. \tag{2.6}$$

In Figure 2.4 we refer to ϑ as the **zenith angle** and to φ as the **azimuth angle**.

2.3 Intensity of Radiation (Radiance)

The term **spectral** and its synonym **monochromatic** ("mono" = one, "chrome" = color) refer to radiation confined to a vanishingly small wavelength interval $d\lambda$ centered about a specified wavelength λ. Thus, the polarized ray in Figure 2.1 represents spectral radiation. The **spectral intensity** $i_\lambda(\lambda, \vartheta, \varphi)$ of a plane source is the power per unit wavelength in the wavelength interval $d\lambda$ centered about wavelength λ, per unit projected area of the source, per unit solid angle, passing in direction (ϑ, φ). Note that the symbol λ appears twice in the notation. This is not

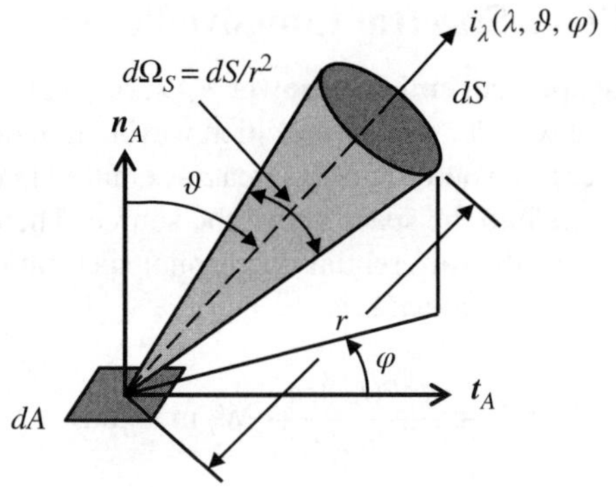

Figure 2.5 A monochromatic beam of spectral intensity $i_\lambda(\lambda, \vartheta, \varphi)$ contained within differential solid angle $d\Omega_s$ leaving a plane area element dA.

redundant usage; the subscript λ signals that the spectral intensity is a per-unit-wavelength quantity, and the λ in the argument list signals that the value of the spectral intensity depends on wavelength. While it is traditional to call this quantity "intensity" in the radiation heat transfer community, it is frequently referred to as "radiance" in the applied optics, astronomy, and earth sciences literature.

Figure 2.5 represents a beam of monochromatic light of spectral power $d^3P(\lambda, \vartheta, \varphi)$ (W) leaving the plane surface element dA in direction (ϑ, φ) at an angle ϑ with respect to the surface normal and contained in a beam whose solid angle is $d\Omega_s$. Then the spectral intensity of this beam is

$$i_\lambda(\lambda, \vartheta, \varphi) = \frac{d^3P(\lambda, \vartheta, \varphi)}{dA \cos\vartheta \, d\Omega_S \, d\lambda} \; (\mathrm{W/m^2 \cdot sr \cdot \mu m}). \tag{2.7}$$

The superscript "3" on the differential operator in the numerator of Eq. (2.7) is required for notational consistency; in order for the intensity to be a finite quantity, the number of differential symbols d must be the same in both the numerator and the denominator.

Another useful concept is the **total intensity** of a beam, which is obtained by integrating the spectral intensity over all wavelengths, i.e.,

$$i(\vartheta, \varphi) = \int_{\lambda=0}^{\infty} i_\lambda(\lambda, \vartheta, \varphi) d\lambda. \tag{2.8}$$

The word "total" is used exclusively in this context in the radiation heat transfer literature.

2.4 Directional Spectral Emissive Power

The **directional spectral emissive power** $e_\lambda(\lambda, \vartheta, \varphi)$ of a plane source is the power per unit wavelength in a specified wavelength interval $d\lambda$ about wavelength λ, per unit source surface area dA, emitted in direction (ϑ, φ) per unit solid angle into the space above the source. Then referring once again to Figure 2.5, the differential directional spectral emissive power contained in the solid angle $d\Omega_s$ is

$$dE_\lambda \equiv \frac{d^3P(\lambda, \vartheta, \varphi)}{dA\, d\lambda}(\text{W/m}^2 \cdot \mu\text{m}). \tag{2.9}$$

Invoking Eq. (2.7) we have

$$dE_\lambda = i_\lambda(\lambda, \vartheta, \phi) \cos \vartheta d\Omega_s\ (\text{W/m}^2 \cdot \mu\text{m}). \tag{2.10}$$

Finally, the directional spectral emissive power is

$$e_\lambda(\lambda, \vartheta, \varphi) \equiv \frac{dE_\lambda}{d\Omega_S} = i_\lambda(\lambda, \vartheta, \phi) \cos \vartheta\ (\text{W/m}^2 \cdot \text{sr} \cdot \mu\text{m}). \tag{2.11}$$

2.5 Hemispherical Spectral Emissive Power

The **hemispherical spectral emissive power** $e_\lambda(\lambda)$ is the power emitted into the hemispherical ("2π") space above a plane source in a specified wavelength interval $d\lambda$ about wavelength λ, per unit source surface area dA, as illustrated in Figure 2.6. The hemispherical spectral emissive power may be expressed in terms of the directional spectral emissive power by integrating over the hemispherical space above the source,

$$e_\lambda(\lambda) = \int_{2\pi} e_\lambda(\lambda, \vartheta, \varphi) d\Omega = \int_{2\pi} i_\lambda(\lambda, \vartheta, \varphi) \cos \vartheta d\Omega \tag{2.12}$$

or

$$e_\lambda(\lambda) = \int_{\varphi=0}^{2\pi} \int_{\vartheta=0}^{\pi/2} i_\lambda(\lambda, \vartheta, \varphi) \cos \vartheta\ \sin \vartheta d\vartheta d\varphi. \tag{2.13}$$

2.6 Hemispherical Total Emissive Power

The **hemispherical total emissive power** e is the power per unit surface area emitted into the hemispherical space above a plane source at all

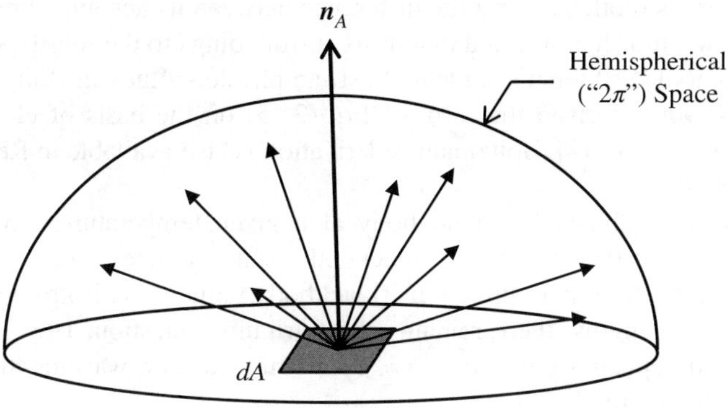

Figure 2.6 Radiation emitted by a surface element *dA* into the overlying hemispherical space.

wavelengths. It is, therefore, related to the spectral intensity according to

$$e = \int_{\lambda=0}^{\infty} \int_{\varphi=0}^{2\pi} \int_{\vartheta=0}^{\pi/2} i_\lambda(\lambda, \vartheta, \varphi) \cos\vartheta \, \sin\vartheta d\vartheta d\varphi d\lambda. \tag{2.14}$$

2.7 The Blackbody Radiation Distribution Function

By definition, a **blackbody** is a perfect absorber of thermal radiation; that is, it absorbs all incident radiation from all directions and at all wavelengths. It follows from this definition and from the Second Law of Thermodynamics that no body at a given temperature can emit more thermal radiation than a blackbody at the same temperature. Therefore, we say that a blackbody is an ideal emitter. Furthermore, it can be demonstrated through the use of "thought experiments" (see Ref. [1], pp. 32–33) that an isothermal enclosure is filled with **blackbody radiation**, which is both uniform and isotropic. According to the Stefan–Boltzmann law, the emissive power escaping from an isothermal enclosure through a vanishingly small hole in the wall is

$$e_b = \sigma T^4 \text{ (W/m}^2\text{)}, \tag{2.15}$$

where T (K) is the absolute temperature and $\sigma = 5.6696 \times 10^{-8}$ W m$^{-2}\cdot$K^4 is the Stefan–Boltzmann constant. The German physicist Josef Stefan (1835–1893) first suggested the form of Eq. (2.15) in 1879 on the basis of data already in the literature (see Problem 2.4) [2]. Stefan discovered that a straight line results when the initial cooling rate of a body suspended in a vacuum is plotted against the difference between its absolute temperature to the fourth power and that of its surroundings to the fourth power. Five years later Stefan's student, Austrian physicist Ludwig Boltzmann (1844–1906), derived the form of Eq. (2.15) on the basis of classical thermodynamics [3]. Boltzmann's derivation is also available in Ref. [1] (pp. 38–42).

Having established that no body at a given temperature can emit more thermal radiation than a blackbody at that temperature, and that this must be true in all directions (blackbody radiation is isotropic) and at all wavelengths, there remains the intriguing question, How is the blackbody spectral intensity $i_{b\lambda}(\lambda, T)$ distributed with wavelength at a given temperature?

As the end of the nineteenth century approached, the best minds of physics were committed to deriving the correct expression for the **blackbody radiation distribution function**, as it came to be called. The consensus view was that, once discovered, it would conform to the following constraints:

1. $i_{b\lambda}(\lambda, T) \longrightarrow 0$ as $\lambda \longrightarrow 0$
2. $i_{b\lambda}(\lambda, T) \longrightarrow 0$ as $\lambda \longrightarrow \infty$
3. $i_{b\lambda}(\lambda, T) \longrightarrow 0$ as $T \longrightarrow 0$
4. $i_{b\lambda}(\lambda, T) \longrightarrow \infty$ as $T \longrightarrow \infty$

The first two constraints are based on the contemporary understanding that EM radiation issues from the simple harmonic motion of charged particles. A vanishingly small wavelength would correspond to an infinite frequency, and thus to a vanishingly small amplitude (due to the speed of light limit), while an infinite wavelength would correspond to a zero frequency of oscillation. Both of these extremes imply a lack of motion. The third and fourth constraints are based on the fundamental tenet of classical mechanics that temperature itself is a measure of motion of elementary particles. Indeed, experimental observations were already available that seemed to confirm these expectations. There only remained to find the mathematical expression that fit the data and for which Eqs. (2.14) and (2.15) produced the same value of emissive power e at a given temperature.

An early result of significant historical interest is that of the German physicist Wilhelm Carl Werner Otto Fritz Franz ("Willy") Wien (1864–1928). In 1896, Wien proposed an expression which, when expressed in terms of intensity, can be written [4]

$$i_{b\lambda}^{W} = C_1 \lambda^{-5} e^{-C_2/\lambda T}, \tag{2.16}$$

where the values of the two "empirical" constants are currently recognized to be $C_1 = 1.191044 \times 10^8$ W μm^4 m^{-2} sr and $C_2 = 14\,388$ μm K. Wien based his semiempirical expression on a combination of classical thermodynamics, intuition, and familiarity with experimental observations available in the contemporary literature, which fit his result very well. However, it does not conform to the fourth constraint listed above. Equation (2.16) is plotted in Figure 2.7.

Lord Rayleigh (John William Strutt, the third Baron Rayleigh, 1842–1919) in 1900 [5] and Sir James Hopwood Jeans (1877–1946) in 1905 [6], working independently, derived another candidate expression for the blackbody radiation distribution function using Maxwell–Boltzmann statistics. At the time this was a radical new statistical view of nature that was thought to correctly describe the consequences of its molecular structure. The Rayleigh–Jeans result, expressed in terms of intensity, is

$$i_{b\lambda}^{R-J} = \frac{C_1\lambda^{-5}}{C_2/\lambda T}, \tag{2.17}$$

where C_1 and C_2 are the same constants as in the Wien expression. Equation (2.17), also plotted in Figure 2.7, is unbounded as the

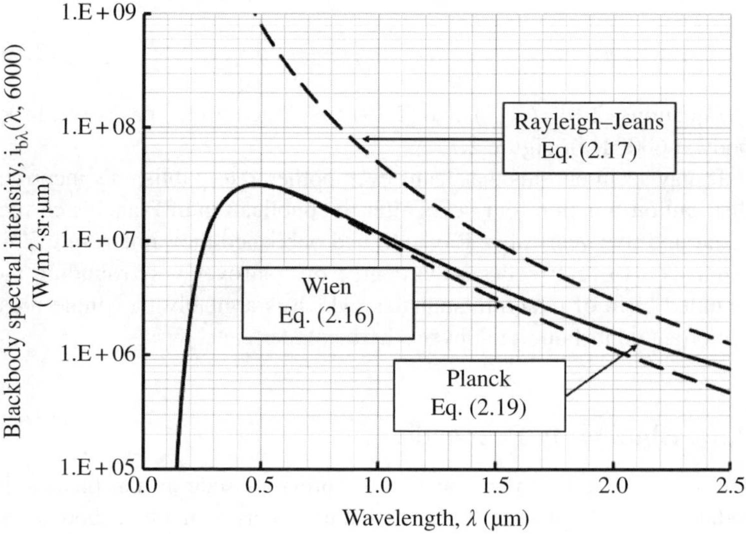

Figure 2.7 Comparison of the Wien, Rayleigh–Jeans, and Planck blackbody radiation distribution functions for a blackbody at a temperature of 6000 K.

wavelength approaches zero, thereby disastrously violating the first physical constraint listed above. The dramatic failure of Maxwell–Boltzmann statistics to correctly describe the spectral distribution of blackbody radiation (deemed the "ultraviolet catastrophe") was seen as a serious setback to physics and prepared the ground for quantum mechanics.

In 1901 the German physicist Max Planck (1858–1947) published an article containing what turned out to be the correct blackbody radiation distribution function [7],

$$i_{b\lambda} = \frac{2c_0^2 h\lambda^{-5}}{e^{hc_0/\lambda kT} - 1}, \tag{2.18}$$

where c_0 is the speed of light in a vacuum, k ($= 1.380 \times 10^{-23}$ J·K^{-1}) is Boltzmann's constant, and h ($= 6.6237 \times 10^{-34}$ J·s) is now known as Planck's constant. The reader interested in the relevant details of the derivation of Eq. (2.18) is referred to Chapter 2 of Ref. [1]. Planck needed the coefficient h, which appeared for the first time in Ref. [7], to fit available data. Therefore, at the time it was considered to be an empirical constant, or "fudge factor." However, it has subsequently joined the speed of light in a vacuum and Boltzmann's constant as one of the fundamental constants of natural science. Equation (2.18) can also be expressed in terms of the constants C_1 and C_2,

$$i_{b\lambda} = \frac{C_1\lambda^{-5}}{e^{C_2/\lambda T} - 1}. \tag{2.19}$$

Equations (2.16), (2.17), and (2.19) are plotted for the case of a blackbody at 6000 K in Figure 2.7.

It may seem curious that Jeans even bothered to publish his incorrect distribution function four years after the publication of Planck's correct result and five years after Rayleigh had published the same result. This sequence of events reflects the degree of controversy surrounding the troubled birth of quantum statistics, which is arguably the single most important contribution to physics in the modern age.

2.8 Blackbody Properties

We learn in the first paragraph in the previous section that blackbody radiation is isotropic. We conclude that the intensity of a blackbody must be independent of direction; that is, for a blackbody

$$i_\lambda(\lambda, \vartheta, \varphi) = i_{b\lambda}(\lambda, T). \tag{2.20}$$

It then follows from Eq. (2.11) that the directional spectral emissive power of a blackbody is

$$e_{b\lambda}(\lambda, T, \vartheta) = i_{b\lambda}(\lambda, T) \cos \vartheta. \tag{2.21}$$

Thus, the directional spectral emissive power of a blackbody varies as the cosine of the angle with respect to the local surface normal. This is sometimes referred to as **Lambert's cosine law**, and surfaces that conform to this law are frequently referred to as **Lambertian**. While all blackbodies are Lambertian, not all Lambertian surfaces are blackbodies.

The blackbody hemispherical spectral emissive power, $e_{b\lambda}(\lambda, T, \vartheta)$, is

$$e_{b\lambda}(\lambda, T) = i_{b\lambda}(\lambda, T) \int_{2\pi} \cos \vartheta d\Omega = \pi i_{b\lambda}(\lambda, T), \tag{2.22}$$

and the blackbody total intensity, $i_b(T)$, is

$$i_b(T) = \int_{\lambda=0}^{\infty} i_{b\lambda}(\lambda, T) d\lambda. \tag{2.23}$$

Evaluation of the integral in Eq. (2.23) is complicated by the form of the integrand, given by Eq. (2.19). The approach is to introduce a change of variables, $\eta = C_2/\lambda T$, after which

$$i_b(T) = \frac{C_1 T^4}{C_2^4} \int_{\eta=0}^{\infty} \frac{\eta^3}{e^\eta - 1} d\eta. \tag{2.24}$$

In the past, evaluation of the improper integral in Eq. (2.24) was a standard homework problem in graduate-level radiation heat transfer courses. However, with the advent of online symbolic integration engines it is now trivial to verify that its value is $\pi^4/15$. Thus, we have

$$i_b(T) = \frac{C_1 \pi^4 T^4}{15 C_2^4} = \frac{1}{\pi} \sigma T^4. \tag{2.25}$$

Finally, the blackbody hemispherical total emissive power is

$$e_b(T) = \int_{\lambda=0}^{\infty} e_{b\lambda}(\lambda, T) d\lambda = \pi \int_{\lambda=0}^{\infty} i_{b\lambda}(\lambda, T) d\lambda = \pi i_b(T) = \sigma T^4. \tag{2.26}$$

The directional distributions of blackbody total intensity and total emissive power are illustrated in Figure 2.8.

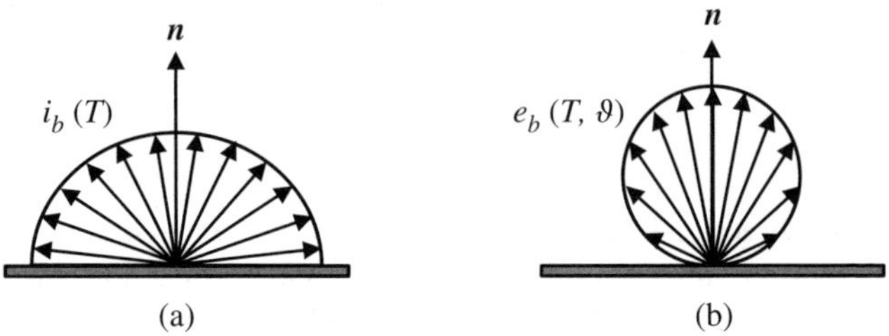

Figure 2.8 Directional distributions of blackbody directional (a) total intensity and (b) total emissivity.

2.9 Emission and Absorption Mechanisms

To this point we have characterized thermal radiation as a wave phenomenon. However, in 1905 Albert Einstein introduced the idea of the photon as an alternative view of EM radiation. In Einstein's photoelectric theory the **photon** is a particle whose energy e (not to be confused with emissive power) is proportional to the frequency of a corresponding EM wave,

$$e = h\upsilon \text{ (J)}, \tag{2.27}$$

where h is Planck's constant. The dual wave-particle description of EM radiation is now firmly established, with one being more convenient to use than the other depending on the situation. Another important conclusion of the photoelectric theory is that, at the most fundamental level, radiation heat transfer always involves interactions between photons and electrons. Modern physics recognizes two categories of atomic particle: the **fermions**, which are the building blocks of matter, and **bosons**, which moderate interactions among the fermions. In the field of quantum electrodynamics (QED), photons and electrons form a boson–fermion pair whose interactions account for all electrical and magnetic phenomena. The reader interested in pursuing this fascinating topic further is referred to Richard P. Feynman's highly readable classic *QED: the Strange Theory of Light and Matter* [8].

For the purposes of the following discussion, an atom may be viewed as a positively charged nucleus surrounded by a swarm of negatively charged electrons. In order for an atom to be electrically neutral, the number of electrons must exactly balance the positive charge of the nucleus. The rules of quantum mechanics require that the electrons organize themselves into layers, or "shells," surrounding the nucleus. A discrete energy level is identified with an electron depending on the shell it occupies, with electrons occupying the inner shells having less energy than those occupying outer shells. Electrons can migrate between shells only by gaining or giving up the amount of energy associated with the difference between their fixed energy in the two shells. The mechanism for gaining or giving up this energy is interaction with a photon, as required by Einstein's photoelectric theory embodied in Eq. (2.27). Thus, when an electron moves from one energy level to another within an atom, the conservation of energy principle requires that a corresponding amount, or **quantum**, of energy be absorbed by or emitted from the atom. If the atomic transition occurs from energy level E_a to a lower energy level E_b, then, according to Eq. (2.27), the frequency of the light emitted by the atom for this **bound-bound transition** is

$$\upsilon = \frac{E_a - E_b}{h}. \tag{2.28}$$

Occasionally a free electron with a speed s will be captured by an atom and come to rest at some energy level, say E_a. In the case of such a **free-bound transition** the frequency of the light emitted will be

$$\upsilon = \frac{1}{h}\left(\frac{1}{2}m_e s^2 - E_a\right). \tag{2.29}$$

Because all speeds s are available to a free electron, Eq. (2.29) permits a **continuum** of frequencies (or wavelengths), whereas Eq. (2.28) permits emission only at discrete frequencies. In the case of high-temperature combustion, thermal plasmas, and electric arcs, a continuum is also produced by the acceleration of free electrons as they follow curved paths. This latter phenomenon is referred to as **bremsstrahlung**, the German word meaning "burning radiation." The transitions represented by Eqs. (2.28) and (2.29), illustrated schematically in Figure 2.9, are completely reversible, with the reverse transitions corresponding to absorption rather than emission.

The atomic transitions described above form the basis for **spectroscopy**, the science that relates the wavelength of radiation emitted from or absorbed by an excited atom to the energy levels occupied by its electrons. Because the energy levels available for occupation are unique to each element, spectroscopy permits identification of the chemical species present in a heated sample. Figure 2.10 is a **spectrogram** showing a portion of the emission spectrum from ionized methane (CH_4) [9].

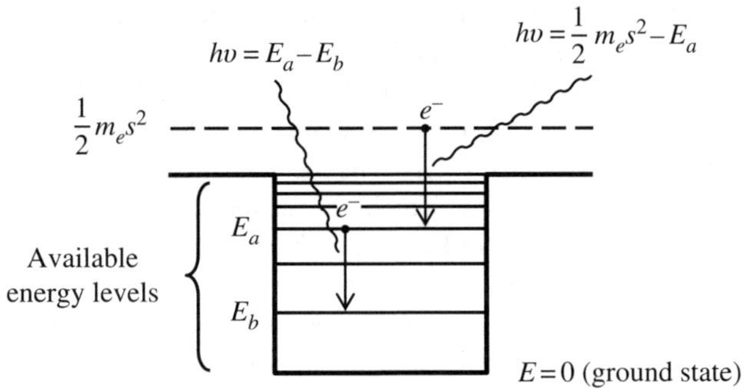

Figure 2.9 The atomic potential well illustrating a bound-bound and a free-bound transition.

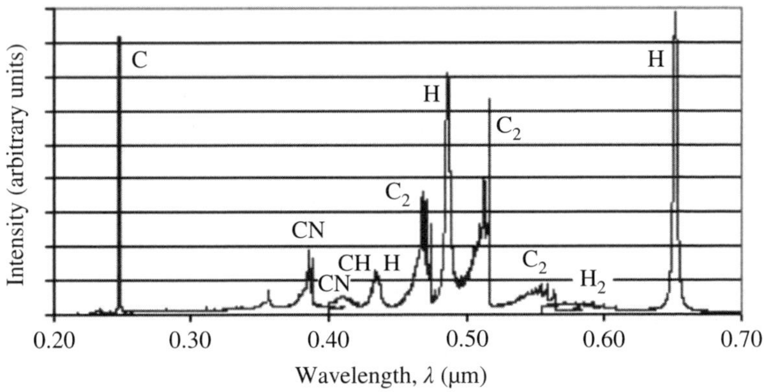

Figure 2.10 The emission spectrum from ionized methane [9].

2.10 Definition of Models for Emission, Absorption, and Reflection

A surface is formally defined in Section 1.4 as the interface between two regions of space having different optical properties, where the optical properties in question are the refractive and absorptive indices n and k. We distinguish between these two **properties** and the **models** used to characterize the interaction between thermal radiation and surfaces. We are now in a position to elaborate on the definition and use of the surface interaction models for emission, reflection, and absorption introduced in Chapter 1.

We define the **directional spectral emissivity** $\varepsilon(\lambda, T, \vartheta, \varphi)$ as the ratio of the spectral intensity of emission from a real body in direction (ϑ, φ) to the spectral intensity of a blackbody at the same temperature,

$$\varepsilon(\lambda, T, \vartheta, \varphi) \equiv \frac{i_{\lambda,e}(\lambda, T, \vartheta, \varphi)}{i_{b\lambda}(\lambda, T)}. \tag{2.30}$$

Note that the symbol for the directional spectral emissivity can also be written $\varepsilon'_{\lambda}(T)$, where the prime (′) indicates directionality and the subscript λ identifies the model as spectral.

The **directional total emissivity** of a surface is then related to the directional spectral emissivity according to

$$\varepsilon'(T) = \varepsilon(T, \vartheta, \varphi) = \frac{\int_{\lambda=0}^{\infty} \varepsilon(\lambda, T, \vartheta, \varphi) i_{b\lambda}(\lambda, T) d\lambda}{\int_{\lambda=0}^{\infty} i_{b\lambda}(\lambda, T) d\lambda}. \tag{2.31}$$

We know that the denominator is $\sigma T^4/\pi$, so Eq. (2.31) can be rewritten

$$\varepsilon'(T) = \varepsilon(T, \vartheta, \varphi) = \frac{\pi}{\sigma T^4} \int_{\lambda=0}^{\infty} \varepsilon(\lambda, T, \vartheta, \varphi) i_{b\lambda}(\lambda, T) d\lambda. \tag{2.32}$$

A surface is said to be **gray** in a given direction (ϑ_1, φ_1) if the directional spectral emissivity is independent of wavelength in that direction. Equation (2.32) then defines a **gray equivalent** directional emissivity for spectral surfaces. The spectral intensities of a blackbody, a graybody, and a hypothetical real surface, all at 6000 K, are compared in Figure 2.11.

The **hemispherical spectral emissivity** of a diffuse surface is then related to the directional spectral emissivity according to

$$\varepsilon_{\lambda}(T) = \varepsilon(\lambda, T) = \frac{\int_{2\pi} \varepsilon(\lambda, T, \vartheta, \varphi) \cos\vartheta d\Omega}{\int_{2\pi} \cos\vartheta d\Omega}. \tag{2.33}$$

We know that the denominator is π, so Eq. (2.33) can be rewritten

$$\varepsilon_{\lambda}(T) = \varepsilon(\lambda, T) = \frac{1}{\pi} \int_{2\pi} \varepsilon(\lambda, T, \vartheta, \varphi) \cos\vartheta d\Omega. \tag{2.34}$$

A surface is said to be **diffuse** at wavelength λ_1 if its directional spectral emissivity is independent of direction at that wavelength. Equation (2.34) then defines a **hemispherical equivalent** spectral emissivity for directional surfaces. The directional, spectral intensities at a specified wavelength, temperature, and azimuth angle for a blackbody,

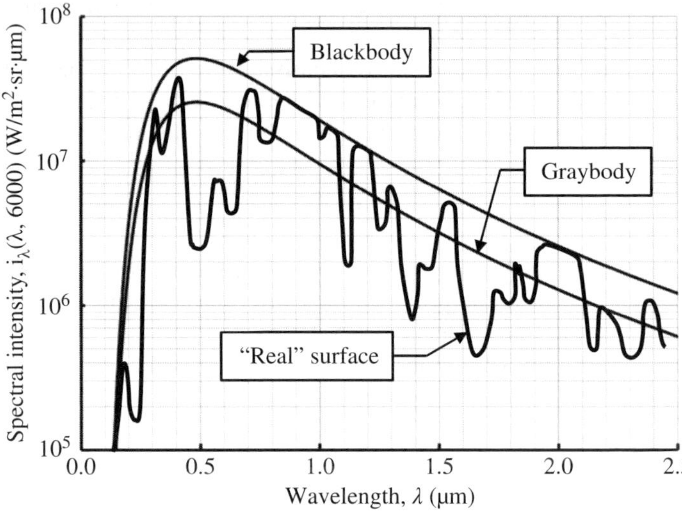

Figure 2.11 Spectral intensity of a blackbody, a graybody, and a hypothetical real surface, all at 6000 K.

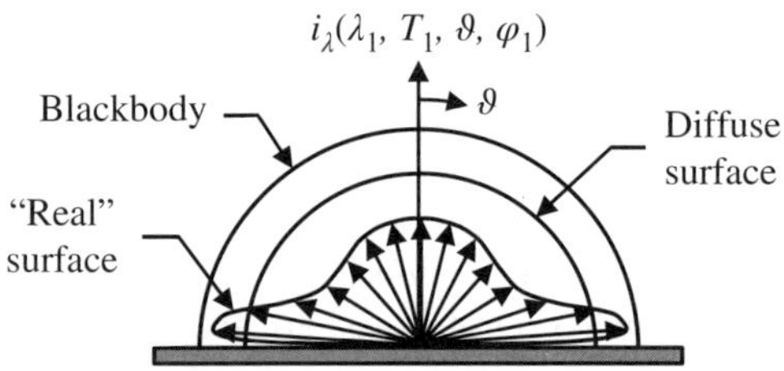

Figure 2.12 The directional spectral intensities at a specified wavelength, temperature, and azimuth angle for a blackbody, a diffuse surface, and a hypothetical real surface.

a diffuse surface, and a hypothetical real surface are compared in Figure 2.12.

The **hemispherical total emissivity** $\varepsilon(T)$ is related to the directional spectral emissivity according to

$$\varepsilon(T) = \frac{1}{\sigma T^4} \int_{\lambda=0}^{\infty} \int_{2\pi} \varepsilon(\lambda, T, \vartheta, \varphi) i_{b\lambda}(\lambda, T) \cos \vartheta d\Omega \, d\lambda. \tag{2.35}$$

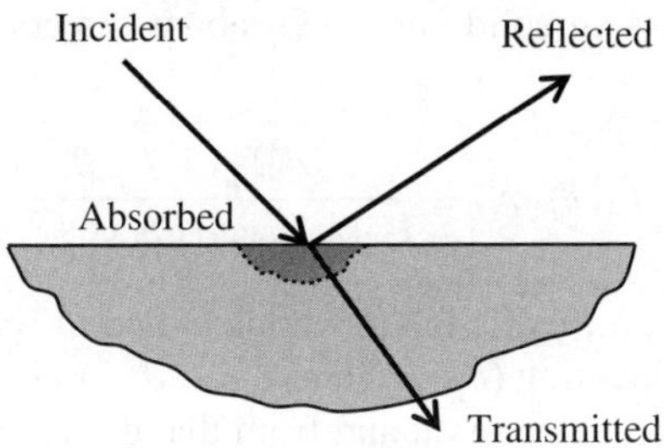

Figure 2.13 The disposition of radiation incident to a surface.

Equation (2.35) then defines an equivalent hemispherical total emissivity for directional spectral surfaces.

When thermal radiation is **incident** to a surface, it is **reflected**, **absorbed**, or **transmitted**, as indicated in Figure 2.13; that is,

$$\text{Incident} = \text{Absorbed} + \text{Reflected} + \text{Transmitted}, \tag{2.36}$$

or

$$1 = \frac{\text{Absorbed}}{\text{Incident}} + \frac{\text{Reflected}}{\text{Incident}} + \frac{\text{Transmitted}}{\text{Incident}}. \tag{2.37}$$

(**Scattering**, which is treated in Chapter 5, is included in transmission for purposes of the current discussion.) The three terms on the right-hand side of Eq. (2.37) are referred to, respectively, as the **absorptivity** α, the **reflectivity** ρ, and the **transmissivity** τ. Thus, Eq. (2.37) can be rewritten

$$1 = \alpha + \rho + \tau. \tag{2.38}$$

A surface or medium is said to be **opaque** if its transmissivity for radiation incident from all directions and at all wavelengths is zero. Then for opaque surfaces,

$$\alpha + \rho = 1 \text{ (opaque surfaces)}, \tag{2.39}$$

or

$$\rho = 1 - \alpha \text{ (opaque surfaces)}. \tag{2.40}$$

Finally, it can be shown that $\alpha = \varepsilon$ under certain circumstances (see Problems 2.9, 2.10, and 2.12).

The **directional spectral absorptivity** $\alpha(\lambda, T_A, \vartheta_i, \varphi_i)$ is the fraction of the power in wavelength interval $d\lambda$ about wavelength λ incident to surface element dA in solid angle $d\Omega_i$ about direction (ϑ_i, φ_i) that is absorbed; that is,

$$\alpha'_\lambda(T_A) = \alpha(\lambda, T_A, \vartheta_i, \varphi_i) \equiv \frac{d^3 P_a(\lambda, T_A, \vartheta_i, \varphi_i)}{i_{\lambda,i}(\lambda, \vartheta_i, \varphi_i)\, dA\, \cos\vartheta_i\, d\Omega_i\, d\lambda}, \tag{2.41}$$

where $d^3P_a(\lambda, T_A, \vartheta_i, \varphi_i)$ is the power absorbed in wavelength interval $d\lambda$ about λ from direction (ϑ_i, φ_i), and $i_{\lambda,i}(\lambda, \vartheta_i, \varphi_i)$ is the intensity incident in that wavelength interval and from that direction. The symbol T_A in Eq. (2.41) and elsewhere in this chapter refers to the temperature of the absorbing surface rather than to the temperature of the irradiating source. The geometry corresponding to this definition is illustrated in Figure 2.14.

In the Monte Carlo ray-trace view of thermal radiation, the directional spectral absorptivity is interpreted as the probability that an energy bundle carrying energy in wavelength interval $\Delta\lambda$ and incident from direction (ϑ_i, φ_i) will be absorbed.

Finally, it can be established through another thought experiment (see Ref. [1], p. 80) involving the Second Law of Thermodynamics that the directional spectral absorptivity is unconditionally equal to the directional spectral emissivity; that is,

$$\alpha(\lambda, T_A, \vartheta_i, \varphi_i) = \varepsilon(\lambda, T_A, \vartheta_e = \vartheta_i, \varphi_e = \varphi_i). \tag{2.42}$$

This result is often referred to as **Kirchhoff's law**.

The **hemispherical spectral absorptivity** $\alpha(\lambda, T_A)$ is the fraction of the power incident to a surface element dA in the wavelength interval $d\lambda$

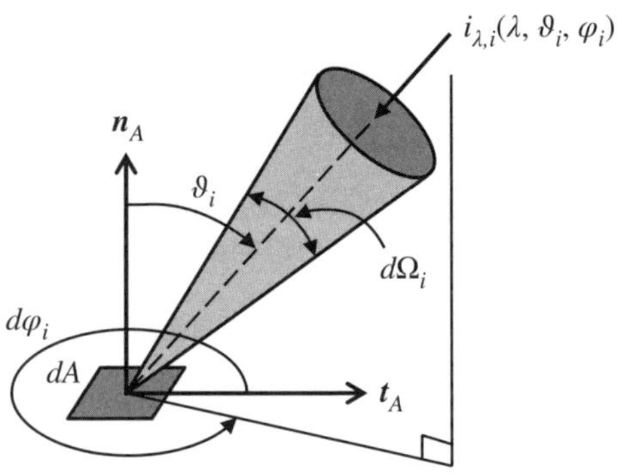

Figure 2.14 A monochromatic beam incident to a surface element.

about λ from the hemispherical space above dA that is absorbed. It can be obtained by integrating the directional spectral absorptivity, weighted by the incident intensity, over the incident hemisphere,

$$\alpha_\lambda(T_A) = \alpha(\lambda, T_A) = \frac{\int_{2\pi} \alpha(\lambda, T_A, \vartheta_i, \phi_i) i_\lambda(\lambda, \vartheta_i, \phi_i) \cos\vartheta_i \, d\Omega_i}{\int_{2\pi} i_\lambda(\lambda, \vartheta_i, \phi_i) \cos\vartheta_i \, d\Omega_i}. \quad (2.43)$$

The **directional total absorptivity** $\alpha(T_A, \vartheta_i, \varphi_i)$ is the total power incident to a surface element dA from direction (ϑ_i, φ_i) in the solid angle $d\Omega_i$ about (ϑ_i, φ_i) that is absorbed. It can be obtained by integrating the directional spectral absorptivity, weighted by the incident intensity, over all wavelengths,

$$\alpha'(T_A) = \alpha(T_A, \vartheta_i, \varphi_i) = \frac{\int_{\lambda=0}^{\infty} \alpha(\lambda, T_A, \vartheta_i, \varphi_i) i_\lambda(\lambda, \vartheta_i, \varphi_i) d\lambda}{\int_{\lambda=0}^{\infty} i_\lambda(\lambda, \vartheta_i, \varphi_i) d\lambda}. \quad (2.44)$$

The **hemispherical total absorptivity** $\alpha(T_A)$ is the total power incident to a surface element dA due to radiation incident from the hemispherical space above dA that is absorbed. It can be obtained by integrating the directional spectral absorptivity, weighted by the incident intensity, over the incident hemisphere and over all wavelengths,

$$\alpha(T_A) = \frac{\int_{\lambda=0}^{\infty} \int_{2\pi} \alpha(\lambda, T_A, \vartheta_i, \varphi_i) i_\lambda(\lambda, \vartheta_i, \varphi_i) \cos\vartheta_i \, d\Omega_i d\lambda}{\int_{\lambda=0}^{\infty} \int_{2\pi} i_\lambda(\lambda, \vartheta_i, \varphi_i) \cos\vartheta_i \, d\Omega_i d\lambda}. \quad (2.45)$$

It should be emphasized here that the radiation incident to surface element dA is generally not black, nor is it gray or diffuse. However, for the important special case of a surface being irradiated by a black source at temperature T_S, Eq. (2.45) can be written

$$\alpha(T_A, T_S) = \frac{1}{\sigma T_S^4} \int_{\lambda=0}^{\infty} \int_{2\pi} \alpha(\lambda, T_A, \vartheta_i, \varphi_i) i_{b\lambda}(\lambda, T_S) \cos\vartheta_i \, d\Omega_i \, d\lambda$$

(black source at T_S). (2.46)

This result is significant because it establishes that the total absorptivity is not, strictly speaking, a model for the surface alone when it is irradiated by a blackbody source; its value in this case also depends on the source temperature. It can be demonstrated (see Problem 2.11) that this is also true if the surface is irradiated with a diffuse gray source.

The **bidirectional spectral reflectivity** determines the intensity $di_{\lambda,r}(\lambda, T_A, \vartheta_i, \varphi_i, \vartheta_r, \varphi_r)$ of a beam reflected from an area element dA due to an incident beam of intensity $i_{\lambda,i}(\lambda, \vartheta_i, \varphi_i)$. Unlike the emissivities and absorptivities defined above, the bidirectional spectral reflectivity is not dimensionless; rather, it is defined as the ratio of the reflected *intensity* to the incident *power*,

$$\rho''_\lambda = \rho(\lambda, T_A, \vartheta_i, \varphi_i, \vartheta_r, \varphi_r) \equiv \frac{di_{\lambda,r}(\lambda, T_A, \vartheta_i, \varphi_i, \vartheta_r, \varphi_r)}{i_{\lambda,i}(\lambda, \vartheta_i, \varphi_i) \cos \vartheta_i d\Omega_i}, \tag{2.47}$$

and, therefore, carries the inverse solid angle unit sr^{-1}. Also, note that while the intensity of the incident beam is finite, the intensity of the reflected beam is an infinitesimal quantity. This is consistent with the fact that, upon reflection, the incident radiation is scattered over 2π-space. However, the incident radiation is confined to an infinitesimal solid angle $d\Omega_i$, and so the bidirectional spectral reflectivity itself remains a finite quantity. Just as the prime ($'$) is used to indicate a directional surface model in notational shorthand, the double prime ($''$) is used to indicate a bidirectional quantity.

In the following paragraphs we define additional reflectivities, all of which can be expressed in terms of the bidirectional spectral reflectivity. These additional reflectivities include the bi-hemispherical total reflectivity, referred to simply as the reflectivity, which appears in Eqs. (2.38) to (2.40). From this we conclude that the bidirectional spectral reflectivity contains all information about the emission, absorption, and reflection of surfaces. Therefore, it is the most valuable of all surface models since, once measured and tabulated, all other surface models can be computed from it. Application of this procedure is the topic of Chapter 4.

An important property of the bidirectional spectral reflectivity is its **reciprocity**,

$$\rho(\lambda, T_A, \vartheta_1, \varphi_1, \vartheta_2, \varphi_2) = \rho(\lambda, T_A, \vartheta_2, \varphi_2, \vartheta_1, \varphi_1), \tag{2.48}$$

where the subscripts 1 and 2 indicate the interchangeable roles of incident and reflected directions.

The **directional-hemispherical spectral reflectivity** $\rho(\lambda, T_A, \vartheta_i, \varphi_i)$, illustrated in Figure 2.15, is the fraction of the spectral power incident to surface element dA from direction (ϑ_i, φ_i) that is reflected into the hemispherical space $\frown_r$ above dA; that is,

$$\rho'_\lambda = \rho(\lambda, T_A, \vartheta_i, \varphi_i) \equiv \frac{\int_{2\pi} di_{\lambda,r}(\lambda, T_A, \vartheta_i, \varphi_i, \vartheta_r, \varphi_r) \cos \vartheta_r d\Omega_r}{i_{\lambda,i}(\lambda, \vartheta_i, \varphi_i) \cos \vartheta_i d\Omega_i}. \tag{2.49}$$

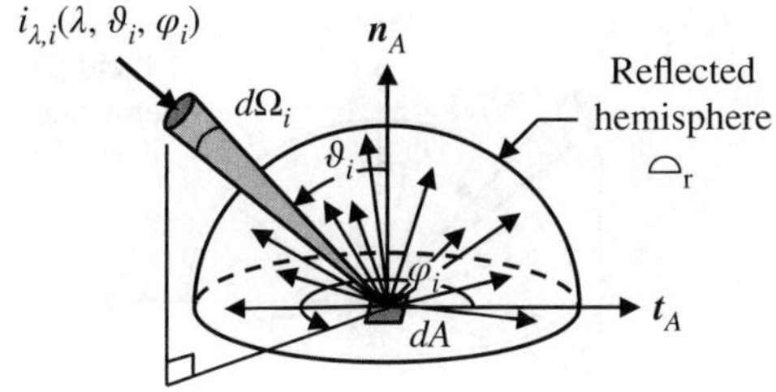

Figure 2.15 Illustration of directional-hemispherical spectral reflectivity.

Rearranging Eq. (2.47) and introducing the result into Eq. (2.49) yields

$$\rho'_\lambda = \rho(\lambda, T_A, \vartheta_i, \varphi_i)$$
$$\equiv \frac{\int_{\triangle_r} \rho''_\lambda(\lambda, T_A, \vartheta_i, \varphi_i, \vartheta_r, \varphi_r) i_{\lambda,i}(\lambda, \vartheta_i, \varphi_i) \cos\vartheta_i d\Omega_i \cos\vartheta_r d\Omega_r}{i_{\lambda,i}(\lambda, \vartheta_i, \varphi_i) \cos\vartheta_i d\Omega_i}, \tag{2.50}$$

where now the 2π limit on the integral in Eq. (2.49) has been replaced with the symbol $\triangle_r$ to emphasize the fact that the integral is over the reflected hemisphere rather than the incident hemisphere. Equation (2.49) may be simplified to

$$\rho'_\lambda = \rho(\lambda, T_A, \vartheta_i, \varphi_i) = \int_{\triangle_r} \rho''_\lambda(\lambda, T_A, \vartheta_i, \varphi_i, \vartheta_r, \varphi_r) \cos\vartheta_r d\Omega_r. \tag{2.51}$$

It is now clear why the bidirectional spectral reflectivity, Eq. (2.47), is defined such that it carries the units of inverse solid angle, sr^{-1}. If this were not so, the directional-hemispherical spectral reflectivity would not be dimensionless.

Finally, recalling Eq. (2.40) for an opaque surface and Kirchhoff's law, Eq. (2.42), we can now write

$$\rho(\lambda, T_A, \vartheta_i, \varphi_i) = 1 - \alpha(\lambda, T_A, \vartheta_i, \varphi_i) = 1 - \varepsilon(\lambda, T_A, \vartheta_i, \varphi_i). \tag{2.52}$$

The significance of the directional-hemispherical spectral reflectivity in the MCRT environment is that it can be interpreted as the probability that an energy bundle incident to the surface will leave behind a fraction $1 - \rho(\lambda, T_A, \vartheta_I, \varphi_i)$ of its energy.

The **hemispherical-directional spectral reflectivity** $\rho(\lambda, T_A, \vartheta_r, \varphi_r)$, illustrated in Figure 2.16, is the fraction of the spectral power incident

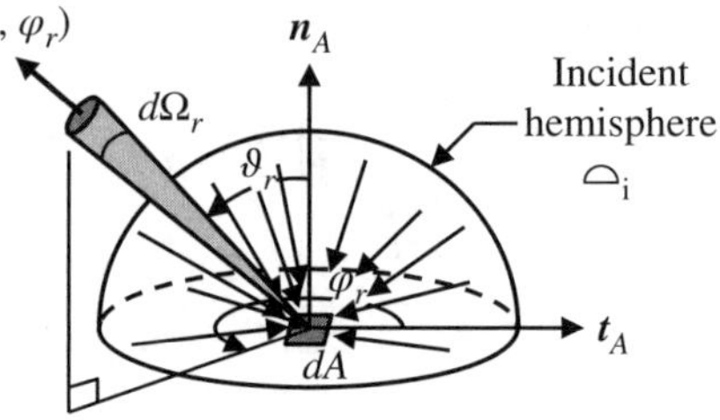

Figure 2.16 Illustration of hemispherical-directional spectral reflectivity.

to surface element dA from the hemispherical space ⌓_i above dA that is reflected as a beam in direction (ϑ_r, φ_r); that is,

$$\rho'_\lambda = \rho(\lambda, T_A, \vartheta_r, \varphi_r) \equiv \frac{i_{\lambda,r}(\lambda, T_A, \vartheta_r, \varphi_r)}{\frac{1}{\pi}\int_{\text{⌓}_i} i_{\lambda,i}(\lambda, \vartheta_i, \varphi_i)\cos\vartheta_i\, d\Omega_i}. \tag{2.53}$$

The factor $1/\pi$ in the denominator of Eq. (2.53) is required for dimensional consistency since the integral is over solid angle. Then in the case of diffuse intensity, $i_{\lambda,i}(\lambda\ only)$, the intensity comes out of the integral and the denominator reduces to $i_{\lambda,i}(\lambda\ only)$; otherwise, the denominator is the mean value of $i_{\lambda,i}(\lambda, \vartheta_i, \varphi_i)$ averaged over the hemisphere.

The numerator in Eq. (2.53) can be replaced by noting that

$$\begin{aligned} i_{\lambda,r}(\lambda, T_A, \vartheta_r, \varphi_r) &= \int_{\text{⌓}_i} di_{\lambda,r}(\lambda, T_A, \vartheta_i, \varphi_i, \vartheta_r, \varphi_r) \\ &= \int_{\text{⌓}_i} \rho(\lambda, T_A, \vartheta_i, \varphi_i, \vartheta_r, \varphi_r) i_{\lambda,i}(\lambda, \vartheta_i, \varphi_i)\cos\vartheta_i\, d\Omega_i \end{aligned} \tag{2.54}$$

where the definition of the bidirectional spectral reflectivity, Eq. (2.47), has been invoked in the second equality. Then introducing Eq. (2.54) into Eq. (2.53) yields

$$\begin{aligned} \rho'_\lambda &= \rho(\lambda, T_A, \vartheta_r, \varphi_r) \\ &\equiv \frac{\int_{\text{⌓}_i} \rho(\lambda, T_A, \vartheta_i, \varphi_i, \vartheta_r, \varphi_r) i_{\lambda,i}(\lambda, \vartheta_i, \varphi_i)\cos\vartheta_i\, d\Omega_i}{\frac{1}{\pi}\int_{\text{⌓}_i} i_{\lambda,i}(\lambda, \vartheta_i, \varphi_i)\cos\vartheta_i\, d\Omega_i}. \end{aligned} \tag{2.55}$$

Inspection of Eq. (2.55) reveals that the hemispherical-directional spectral reflectivity is the intensity-weighted mean, averaged over the incident hemisphere, of the bidirectional spectral reflectivity. Therefore, when the incident radiation is isotropic (diffuse source field), Eq. (2.55) becomes

$$\rho(\lambda, T_A, \vartheta_r, \varphi_r) = \int_{\Omega_i} \rho(\lambda, T_A, \vartheta_i, \varphi_i, \vartheta_r, \varphi_r) \cos \vartheta_i \, d\Omega_i$$

(isotropic radiation). (2.56)

It follows that in this special case

$$\rho(\lambda, T_A, \vartheta_r, \varphi_r) = \rho(\lambda, T_A, \vartheta_i, \varphi_i) \text{ (isotropic radiation).} \quad (2.57)$$

The **(bi)hemispherical spectral reflectivity** $\rho(\lambda, T_A)$, illustrated in Figure 2.17, is the fraction of the spectral power incident to dA from the hemispherical space Ω_i above dA that is reflected back into that hemispherical space Ω_r; that is,

$$\rho(\lambda, T_A) \equiv \frac{\int_{\Omega_r} i_{\lambda,r}(\lambda, T_A, \vartheta_r, \varphi_r) \cos \vartheta_r \, d\Omega_r}{\int_{\Omega_i} i_{\lambda,i}(\lambda, \vartheta_i, \varphi_i) \cos \vartheta_i \, d\Omega_i}. \quad (2.58)$$

Following a procedure similar to that leading to Eq. (2.55), Eq. (2.58) becomes

$$\rho(\lambda, T_A) = \frac{\int_{\Omega_r} \int_{\Omega_i} \rho(\lambda, T_A, \vartheta_i, \varphi_i, \vartheta_r, \varphi_r) i_{\lambda,i}(\lambda, \vartheta_i, \varphi_i) \cos \vartheta_i \, d\Omega_i \cos \vartheta_r \, d\Omega_r}{\int_{\Omega_i} i_{\lambda,i}(\lambda, \vartheta_i, \varphi_i) \cos \vartheta_i \, d\Omega_i}, \quad (2.59)$$

or

$$\rho(\lambda, T_A) = \frac{\int_{\Omega_i} \rho(\lambda, T_A, \vartheta_i, \varphi_i) i_{\lambda,i}(\lambda, \vartheta_i, \varphi_i) \cos \vartheta_i \, d\Omega_i}{\int_{\Omega_i} i_{\lambda,i}(\lambda, \vartheta_i, \varphi_i) \cos \vartheta_i \, d\Omega_i}. \quad (2.60)$$

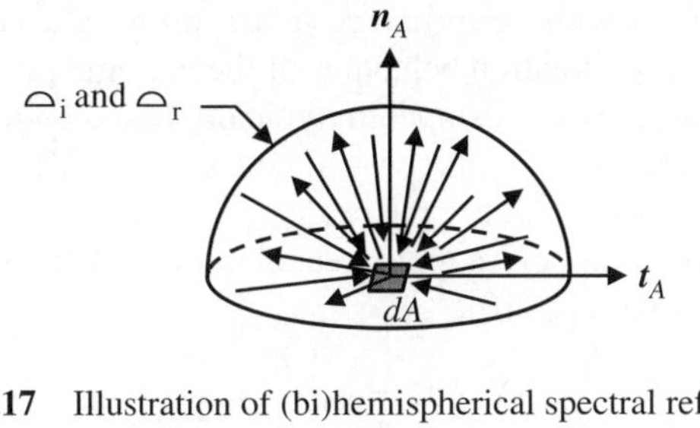

Figure 2.17 Illustration of (bi)hemispherical spectral reflectivity.

The "totalizing" procedure for reflectivities is illustrated by considering the **directional-hemispherical total reflectivity**, $\rho(T_A, \vartheta_i, \varphi_i)$, defined as the fraction of the total power contained in a beam of intensity $i(\vartheta_i, \varphi_i)$ incident to surface element dA that is reflected into the hemispherical space $\triangle_r$ above dA; that is,

$$\rho(T_A, \vartheta_i, \varphi_i) = \frac{\int_{\lambda=0}^{\infty} \rho(\lambda, T_A, \vartheta_i, \varphi_i) i_{\lambda,i}(\lambda, \vartheta_i, \varphi_i) d\lambda_i}{\int_{\lambda=0}^{\infty} i_{\lambda,i}(\lambda, \vartheta_i, \varphi_i) d\lambda_i}. \quad (2.61)$$

This same procedure can be used to express the other total reflectivities in terms of the corresponding spectral reflectivities; that is, the total reflectivity in each case is the weighted average of the corresponding spectral reflectivity where the weight factor is the incident spectral intensity.

2.11 Introduction to the Radiation Behavior of Surfaces

The primary surface conditions that influence the radiation behavior of a solid surface are its bulk electrical properties (electrical conductor or non-conductor), its topography (smooth, polished, sanded, sand blasted, turned, lapped, honed, ground, peened, etc.), its chemical condition (reduced, oxidized, anodized, galvanized, etc.), its degree of contamination (clean or dirty, dusty, dry or oily, etc.), and its surface grain structure (annealed, cold rolled, hot rolled, etc.). Surfaces may also be painted, sputter coated, or evaporation coated to enhance or diminish emission, absorption, or reflection or to bias directionality and/or wavelength dependence. In addition, all surface preparations are subject to damage and aging. The result is a bewilderingly subjective array of adjectives, often open to interpretation, which renders effective communication between designers and modelers difficult if not impossible. Still, it is imperative, once a project moves out of the preliminary design phase, that engineers charged with performance modeling have access to accurate models for surface radiation behavior. In the extreme this often requires a surface characterization campaign, typically based on measurement of the bidirectional spectral reflectivity of key surfaces. The following brief review is intended as a guide to the reader tasked with formulating surface radiation behavior models, a topic treated in more detail in Chapter 4.

Maxwell's Equations [10],

$$\nabla \times \boldsymbol{H} = \varepsilon_0 \frac{\partial \boldsymbol{E}}{\partial t} + \frac{\boldsymbol{E}}{r_e} \quad (2.62)$$

$$\nabla \times \boldsymbol{E} = -\mu_0 \frac{\partial \boldsymbol{H}}{\partial t} \tag{2.63}$$

$$\nabla \cdot \boldsymbol{E} = \frac{\rho_e}{\varepsilon} \tag{2.64}$$

$$\nabla \cdot \boldsymbol{H} = 0, \tag{2.65}$$

are the point of departure for understanding the interaction of EM radiation with a surface. In Eqs. (2.62) to (2.65), $\boldsymbol{H}$ ($\mathrm{A\,m^{-1}}$) is the **magnetic field strength**, E ($\mathrm{V\,m^{-1}}$) is the **electric field strength**, $\varepsilon_0 = 8.854\ \mathrm{C^2/N{\cdot}m^2}$ is the **permittivity of free space**, r_e (Ω-m) is the **electrical resistivity**, ρ_e ($\mathrm{C\,m^{-3}}$) is the **electric charge density**, and $\mu_0 = 4\pi \times 10^{-7}\ \mathrm{N\,A^{-2}}$ is the **magnetic permeability**. These celebrated equations were formulated in 1864 by the British physicist and mathematician James Clerk Maxwell, who synthesized them from already known relationships between electricity and magnetism. Their solution permitted for the first time the theoretical calculation (see Problem 2.7) of the already known speed of light in a vacuum, thereby removing any doubt as to their validity.

Equation (2.1) describes the electric component of EM radiation propagating in free space, without attenuation. We now consider propagation through a homogeneous, isotropic electrical conductor. In this case the term $\mathbf{E}/\mathrm{r_e}$ in Eq. (2.62) representing a **current density**

$$\boldsymbol{J} = \boldsymbol{E}/r_e\ (\mathrm{A/m^2}) \tag{2.66}$$

is not negligible. When electric current flows through a resistance, electrical energy is converted irreversibly into sensible heat according to

$$\dot{q} = \boldsymbol{J} \cdot \boldsymbol{E} = \frac{\boldsymbol{E}}{r_e} \cdot \boldsymbol{E} = \frac{|\boldsymbol{E}|^2}{r_e}\ (\mathrm{W/m^3}). \tag{2.67}$$

Thus, as an electromagnetic wave propagates through a conducting medium it loses power; that is, some of its power is left behind as sensible heat in the electrically conducting medium through which it passes. This is the macroscopic view of the mechanism of absorption of radiation within a material substance.

In the example of the x-propagating, y-polarized, electromagnetic wave illustrated in Figure 2.1, absorption may be accounted for heuristically by multiplying Eq. (2.1) by an exponential attenuation factor; that is,

$$E_y = \left[|E_y| e^{i\omega\left(\frac{nx}{c_0} - t\right)} \right] e^{-\omega kx/c_0}. \tag{2.68}$$

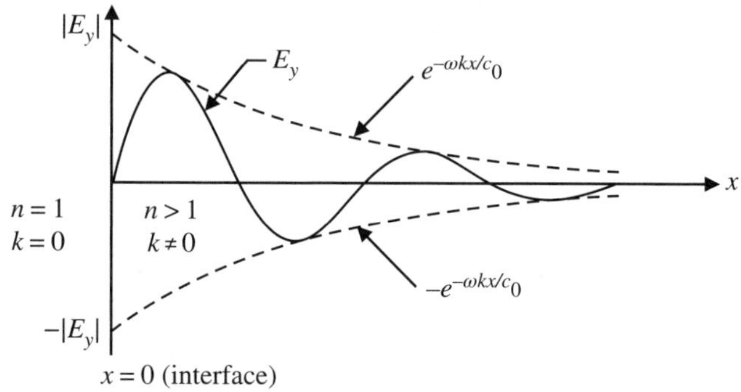

Figure 2.18 Extinction of the electric component of a y-polarized EM wave propagating in the positive-x direction through an electrical conductor.

In Equation (2.68) k is the **extinction coefficient**, also referred to as the **absorption index**, of the medium. It can be demonstrated by direct substitution that Eq. (2.68) satisfies Maxwell's equations with r_e finite if

$$n^2 = c_0^2 \mu \varepsilon = (n + ik)^2 - \frac{i\mu\lambda_0 c_0}{2\pi r_e}, \tag{2.69}$$

where $\lambda_0 = c_0/\nu$. Figure 2.18 illustrates the progressive extinction of the electric component of an x-propagating, y-polarized EM wave as it passes into and through an absorbing medium. The figure clearly demonstrates that absorption is not strictly speaking a "surface" phenomenon.

2.12 Radiation Behavior of Surfaces Composed of Electrical Non-Conductors (Dielectrics)

In the following paragraphs we consider the two limiting cases in which a monochromatic EM wave is incident to the plane interface separating two ideal regions. In the first extreme, the wave passes from a **dielectric** whose optical properties are n_1 and $k_1 = 0$ ($r_e \to \infty$) into another dielectric whose optical properties are n_2 and $k_2 = 0$; in the second extreme, the wave passes from a dielectric whose optical properties are n_1 and $k_1 = 0$ into an electrical conductor, or metal, whose optical properties are n_2 and $k_2 \neq 0$.

The case of a smooth plane interface between two dielectrics with $n_2 > n_1$ is illustrated in Figure 2.19. In the figure $\boldsymbol{E}_{p,i}$ represents the

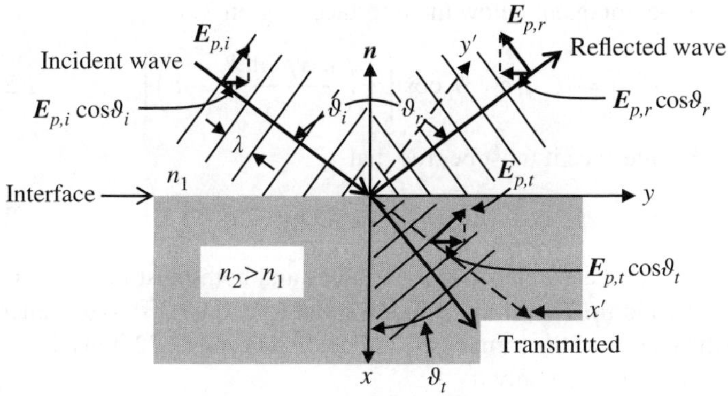

Figure 2.19 A transverse-magnetic (TM), p-polarized, monochromatic wave incident to a smooth plane interface between two dielectrics with $n_2 > n_1$.

electric component of a **transverse-magnetic (TM), p-polarized**, monochromatic electromagnetic wave incident to the interface, or surface. The arrows labeled "Incident," "Reflected," and "Transmitted" can be thought of as "rays," in which case the lines passing normal to the rays indicate wavefronts separated by a distance λ. Following convention, the subscript "p" is used to remind us that the electric field vector in this case lies in the **plane of incidence**, the plane containing both the incident ray and the unit normal vector $\boldsymbol{n} = -\boldsymbol{i}$. Without loss of generality we consider only the real part of the incident electric field,

$$Re[E_{p,i}] = |E_{p,i}| \cos\left[\omega\left(\frac{n_1 x'}{c_0} - t\right)\right], \tag{2.70}$$

where $\omega = 2\pi c_0/\lambda$ and $x' = y/\sin\vartheta_i$. This is equivalent to assuming that the phase angle of the incident wave, $\phi_i = \tan^{-1}[Im(E_{p,i})/Re(E_{p,i})]$, is zero.

Careful consideration of Figure 2.19 reveals that the y-component (parallel to the interface) of the electric field above the interface (region 1) is

$$|E_{y,1}| = |E_{p,i}|\cos\vartheta_i \cos\left[\omega\left(\frac{n_1 y/\sin\vartheta_i}{c_0} - t\right)\right]$$
$$- |E_{p,r}|\cos\vartheta_r \cos\left[\omega\left(\frac{n_1 y/\sin\vartheta_r}{c_0} - t\right)\right] \tag{2.71}$$

and the y-component below the interface (region 2) is

$$|E_{y,2}| = |E_{p,t}| \cos \vartheta_t \cos \left[\omega \left(\frac{n_2 y/ \sin \vartheta_t}{c_0} - t\right)\right]. \tag{2.72}$$

At the interface it must be true that

$$|E_{y,2}| = |E_{y,1}| \quad (x = 0). \tag{2.73}$$

Now y and t are independent variables and both must be independent of ϑ_i, ϑ_r, and ϑ_t. Therefore, it follows that for Eq. (2.73) to be generally true the factors of the form cos[·] in Eqs. (2.71) and (2.72) must be equal. But this can be true only if

$$n_1 \sin \vartheta_i = n_1 \sin \vartheta_r = n_2 \sin \vartheta_t. \tag{2.74}$$

Equation (2.74) leads directly to the **law of specular reflection**

$$\vartheta_r = \vartheta_i \tag{2.75}$$

and the **Snell–Descartes law**,

$$\frac{\sin \vartheta_i}{\sin \vartheta_t} = \frac{n_2}{n_1}. \tag{2.76}$$

Equations (2.71)–(2.76) can be combined to yield

$$|E_{p,i}| \cos \vartheta - |E_{p,r}| \cos \vartheta = |E_{p,t}| \cos \vartheta_t, \tag{2.77}$$

where now $\vartheta = \vartheta_i = \vartheta_r$.

The magnetic field strength vector **H** is perpendicular to the electric field strength vector **E**, as indicated in Figure 2.1. Therefore, the magnetic field strength vector associated with the incident electric field $E_{p,i}$ is perpendicular to the plane of incidence. This perpendicular component of the magnetic field strength is also continuous at the interface; that is,

$$|H_{p,i}| + |H_{p,r}| = |H_{p,t}| \quad (x = 0). \tag{2.78}$$

In writing Eq. (2.78) it has been recognized that, as in Eqs. (2.71) and (2.72), the cos[·] factors are all equal and so have divided out. Also, since the transverse-magnetic field is already parallel to the interface, the cos ϑ factors that appear in Eq. (2.77) are all equal to unity in Eq. (2.78). Like E_p, the quantity H_p is in general a complex quantity. However, H_p is in phase with E_p at incidence and so can also be considered real without loss of generality.

Now, for the special case of the p-polarized EM wave propagating in the positive-x' direction, Eq. (2.63) becomes

$$\frac{\partial E_p}{\partial x'} = -\mu \frac{\partial H_p}{\partial t}, \tag{2.79}$$

where

$$E_p = |E_p| e^{i\omega\left(\frac{x'}{c} - t\right)} \tag{2.80}$$

and

$$H_p = |H_p| e^{i\omega\left(\frac{x'}{c} - t\right)}. \tag{2.81}$$

With the introduction of Eqs. (2.80) and (2.81), Eq. (2.79) becomes

$$|H_p| = \frac{n}{\mu c_0} |E_p|. \tag{2.82}$$

The magnetic permeability μ for most dielectrics is sufficiently near the free-space value μ_0 to be considered invariant across the interface. Subject to this assumption, substitution of Eq. (2.82) into Eq. (2.78) yields

$$n_1 |E_{p,i}| + n_1 |E_{p,r}| = n_2 |E_{p,t}| \quad (x = 0). \tag{2.83}$$

Finally, eliminating $|E_{p,\,t}|$ between Eqs. (2.77) and (2.83) yields

$$\frac{|E_{p,r}|}{|E_{p,i}|} = \frac{\cos\vartheta / \cos\vartheta_t - n_1/n_2}{\cos\vartheta / \cos\vartheta_t + n_1/n_2}, \tag{2.84}$$

which becomes, after introducing the Snell–Descartes law, Eq. (2.76),

$$\frac{|E_{p,r}|}{|E_{p,i}|} = \frac{\tan(\vartheta - \vartheta_t)}{\tan(\vartheta + \vartheta_t)}. \tag{2.85}$$

Natural (unpolarized) radiation can be resolved into two orthogonal EM wave components (see p. 184). An analysis similar to that leading to Eq. (2.85) can be carried out for an incident **transverse-electric (TE)**, **s-polarized** EM wave whose electric field strength vector is perpendicular to the plane of incidence. This leads to

$$\frac{|E_{s,r}|}{|E_{s,i}|} = \frac{\sin(\vartheta - \vartheta_t)}{\sin(\vartheta + \vartheta_t)}, \tag{2.86}$$

where now, following convention, the subscript "s" indicates a TE-polarized wave.

The ratio of the reflected power to the incident power for a p-polarized wave is

$$\rho'_{p,\lambda} = \left(\frac{|E_{p,r}|}{|E_{p,i}|}\right)^2 \tag{2.87}$$

and for an s-polarized wave is

$$\rho'_{s,\lambda} = \left(\frac{|E_{s,r}|}{|E_{s,i}|}\right)^2. \tag{2.88}$$

Note that, even though they involve both an incident direction and a known reflected direction, as established by Eq. (2.75), Eqs. (2.87) and (2.88) are *not* bidirectional reflectivities as defined by Eq. (2.47). We refer to them as directional-hemispherical reflectivities because they conform to the definition preceding Eq. (2.49); that is, they represent the fraction of the power incident to surface element dA from direction (ϑ_i, φ_i) that is reflected into the hemispherical space above dA. In fact, the usual definition of bidirectional reflectivity is not applicable to perfectly specular surfaces. Equations (2.87) and (2.88) are spectral reflectivities because the refractive index n is wavelength dependent.

Finally, for naturally polarized radiation we can write

$$\begin{aligned}\rho'_\lambda &= \frac{1}{2}(\rho'_{p,\lambda} + \rho'_{s,\lambda}) = \frac{1}{2}\left[\frac{\tan^2(\vartheta - \vartheta_t)}{\tan^2(\vartheta + \vartheta_t)} + \frac{\sin^2(\vartheta - \vartheta_t)}{\sin^2(\vartheta + \vartheta_t)}\right] \\ &= \frac{1}{2}\left[\frac{\sin^2(\vartheta - \vartheta_t)}{\sin^2(\vartheta + \vartheta_t)}\right]\left[1 + \frac{\cos^2(\vartheta + \vartheta_t)}{\cos^2(\vartheta - \vartheta_t)}\right] \text{ (dielectrics).}\end{aligned} \tag{2.89}$$

This result is known as **Fresnel's equation**. Note that the Snell–Descartes law, Eq. (2.76), can be used to eliminate ϑ_t from Eq. (2.89) in favor of n_1/n_2.

As an example of the utility of Fresnel's equation, we consider opaque dielectrics such as certain ceramics and stained glasses for which the assumption of $k \cong 0$ invoked in the derivation of Fresnel's equation still permits absorption in a sufficiently thick sample. In this case Kirchhoff's law, Eq. (2.42), together with the definition of an opaque material, Eq. (2.40), allows us to write

$$\varepsilon'_\lambda = \alpha'_\lambda = 1 - \rho'_\lambda. \tag{2.90}$$

Figure 2.20 shows the directional spectral emissivity (= absorptivity) for black glass at a wavelength of 0.546 μm. The symbols are measured

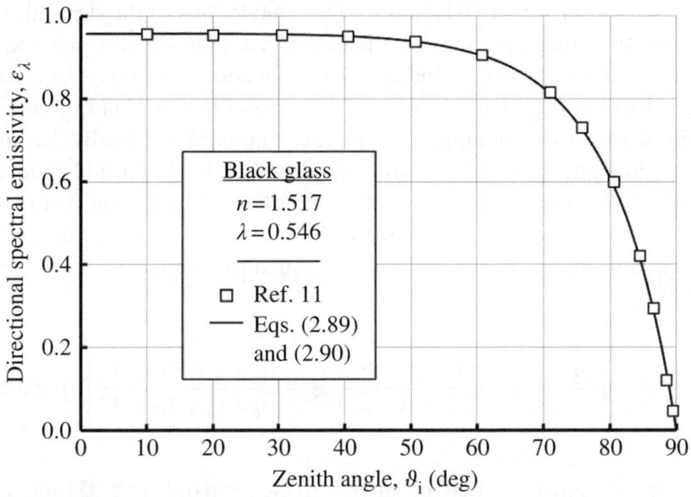

Figure 2.20 Comparison of theory and experiment for the directional spectral emissivity of black glass (n = 1.517) at 0.546 μm.

values corresponding to a refractive index of 1.517 [11], and the curve is based on Eqs. (2.89) and (2.90). The excellent agreement between theory and experiment in this example is gratifying. The behavior of black glass exhibited in Figure 2.20 is typical for electrical non-conductors in that (a) they tend to be efficient emitters and absorbers of thermal radiation, and (b) they emit more efficiently in directions near the normal than they do for grazing angles.

2.13 Radiation Behavior of Surfaces Composed of Electrical Conductors (Metals)

A successful model is also available for the radiation behavior of electrical conductors (metals),

$$\rho'_{p,\lambda} = \frac{(n\cos\vartheta - 1)^2 + (k\cos\vartheta)^2}{(n\cos\vartheta + 1)^2 + (k\cos\vartheta)^2} \tag{2.91}$$

and

$$\rho'_{s,\lambda} = \frac{(n - \cos\vartheta)^2 + k^2}{(n + \cos\vartheta)^2 + k^2}. \tag{2.92}$$

Derivation of Eqs. (2.91) and (2.92) is analogous to the derivation of the corresponding equations for dielectrics, but now we must include the influence of the absorption index k. The derivation is based on Eq. (2.68) rather than Eq. (2.70) and involves the process illustrated in Figure 2.18. Along the way we encounter a complex version of the Snell–Descartes law whose physical interpretation is elusive. The interested reader is referred to Section 4.5 of Ref. [1] for an outline of the derivation of these equations. Finally, Eqs. (2.91) and (2.92), along with Eq. (2.90) and the first equality in Eq. (2.89), may be combined to obtain

$$\varepsilon'_\lambda = \alpha'_\lambda = 1 - \frac{1}{2}\left[\frac{(n\cos\vartheta - 1)^2 + (k\cos\vartheta)^2}{(n\cos\vartheta + 1)^2 + (k\cos\vartheta)^2} + \frac{(n - \cos\vartheta)^2 + k^2}{(n + \cos\vartheta)^2 + k^2}\right] \text{ (metals).} \tag{2.93}$$

A direct comparison between theory in the form of Eq. (2.93) and measured values of the directional spectral emissivity for pure platinum [11] is shown in Figure 2.21.

Fresnel theory for dielectrics and its extension to metals generally provide excellent agreement with laboratory measurements as long as adequate care is taken to ensure that the samples are pure, clean, and

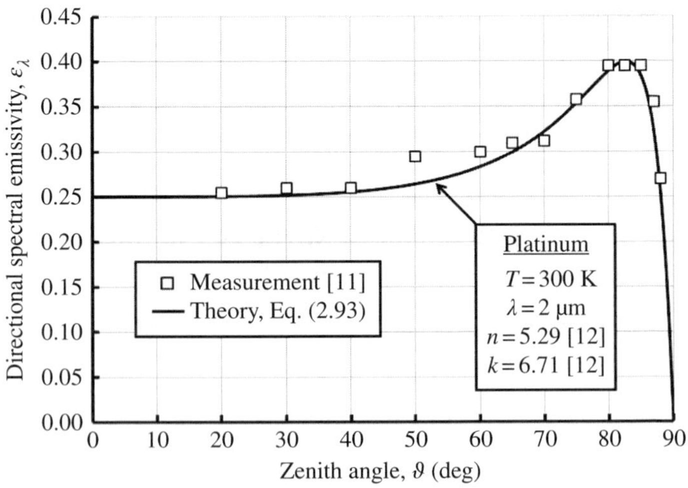

Figure 2.21 Comparison of theory and experiment for the directional spectral emissivity of pure platinum (n = 5.29, k = 6.71) at 2.0 μm.

extremely smooth. Therefore, the results presented in this section and Section 2.12 are adequate for generalizing the optical behavior of metals and non-metals. Comparison of Figures 2.20 and 2.21 makes it clear that (i) metals are generally less efficient emitters and absorbers than dielectrics, while they are more efficient reflectors; and (ii) metals, in sharp contrast to non-metals, emit and absorb more efficiently at grazing angles than near the normal.

Increased surface roughness, aging, and contamination produce significant deviations from the ideal behavior predicted by Fresnel theory, and, of course, oxidation of metals and the purposeful addition of surface coatings to both metal and non-metal surfaces greatly influence their optical behavior. Still, the models presented here are useful from both the pedagogical and practical points of view. For example, they can be "hemisphericalized" and "totalized" as indicated in Section 2.10, and the resulting hemispherical total models used with the material presented in Chapter 3 to carry out conceptual and preliminary design based on the diffuse gray assumption. However, detailed design and serious performance analysis must await the treatment presented in Chapter 4 for enclosures composed of directional spectral surfaces.

Problems

2.1 Demonstrate by integration of Eq. (2.6) that the solid angle subtended by a hemisphere at its center is 2π, and thus that the solid angle subtended by a sphere at its center is 4π.

2.2 Argue that the intensity of a (spherical) point source is independent of (i) direction and (ii) distance from the source. Hint: Consider expressions for the intensity at two locations a distance r_1 and r_2 from the source.

2.3 The **blackbody function** is defined

$$\frac{e_{b\lambda(\lambda,T)}}{T^5} \equiv \frac{2\pi c_0^2 h}{(\lambda T)^5}\left(\frac{1}{e^{hc_0/k\lambda T}-1}\right) = f(\lambda T\ only)\ (\mathrm{W\,m^{-2}\cdot \mu m\cdot K^5}). \tag{2.94}$$

(a) Use Excel or Matlab to plot the blackbody function for $0 \le \lambda T \le 25\,000\ \mu\mathrm{m{\cdot}K}$.

(b) Find the value of λT at the peak of the blackbody function. Hint: find the value of λT for which $\partial f(\lambda T)/\partial(\lambda T) = 0$.

2.4 In a classical article published by Dulong and Petit in 1817 [13], the authors describe an experiment in which a small body is suspended in a bell jar whose outer walls were encased in melting ice. During the experiment, which was designed to measure the thermal capacity of the material from which the body was fabricated, the body was allowed to cool from an initial temperature and its temperature decrease after one minute was recorded. The results are given in Table 2.1. Stefan used the same data in 1879 (62 years later!) to establish the absolute-temperature-to-the-fourth-power dependence of radiant heat emission from a solid body [2]. Can you reproduce Stefan's reasoning and results?

Table 2.1 Data from Dulong and Petit [13] used by Stefan [2] to establish the absolute-temperature-to-the-fourth-power dependence of radiant heat emission from a solid body.

Initial temperature of body (°C)	Temperature decrease after 1 min (°C)
240	10.69
220	8.92
200	7.40
180	6.10
160	4.89
140	3.88
120	3.02
100	2.30
80	1.74

2.5 Demonstrate that the Planck blackbody radiation distribution function, Eq. (2.19), converges to the Wien blackbody radiation distribution function, Eq. (2.16), for sufficiently short wavelengths.

2.6 Demonstrate that the Planck blackbody radiation distribution function, Eq. (2.19), converges to the Rayleigh–Jeans blackbody radiation distribution function, Eq. (2.17), for sufficiently long wavelengths.

2.7 Maxwell's equations describing the propagation of an EM wave through a vacuum may be written

$$\nabla \times \boldsymbol{H} = \varepsilon_0 \frac{\partial \boldsymbol{E}}{\partial t} \tag{2.95}$$

$$\nabla \times \boldsymbol{E} = -\mu_0 \frac{\partial \boldsymbol{H}}{\partial t} \tag{2.96}$$

$$\nabla \cdot \boldsymbol{E} = 0 \tag{2.97}$$

$$\nabla \cdot \boldsymbol{H} = 0, \tag{2.98}$$

where $\boldsymbol{H}$ (A m^{-1}) is the **magnetic field strength**, E (V m^{-1}) is the **electric field strength**, $\varepsilon_0 = 8.854$ C^2/N·m^2 is the **permittivity of free space**, and $\mu_0 = 4\pi \times 10^{-7}$ N A^{-2} is the **magnetic permeability**.

(a) Demonstrate that when an EM wave is propagating in a vacuum in the x direction, Maxwell's equations can be written in Cartesian coordinates

$$\mu\varepsilon \frac{\partial^2 E_y}{\partial t^2} = \frac{\partial^2 E_y}{\partial x^2} \tag{2.99}$$

$$\mu\varepsilon \frac{\partial^2 E_z}{\partial t^2} = \frac{\partial^2 E_z}{\partial x^2} \tag{2.100}$$

$$\mu\varepsilon \frac{\partial^2 H_y}{\partial t^2} = \frac{\partial^2 H_y}{\partial x^2} \tag{2.101}$$

$$\mu\varepsilon \frac{\partial^2 H_z}{\partial t^2} = \frac{\partial^2 H_z}{\partial x^2} \tag{2.102}$$

(b) Demonstrate that $E_y = f(x\sqrt{\mu\varepsilon} - t) + g(x\sqrt{\mu\varepsilon} + t)$, where f and g are any single-valued, twice differentiable functions of x and t, is a solution to Eq. (2.99).

(c) Demonstrate that the function f in Part (b) represents an electromagnetic wave propagating in the positive-x direction at the speed of light in a vacuum, $c_0 = \frac{1}{\sqrt{\mu_0\varepsilon_0}} = 2.997925 \times 10^8$ m s^{-1}.

2.8 Evaluate the integral $\int_{2\pi} \cos \vartheta \, d\Omega$.

2.9 Show that the hemispherical spectral absorptivity is equal to the hemispherical spectral emissivity for the special case of a diffuse surface.

2.10 Show that the directional total absorptivity is equal to the directional total emissivity for the special case of a "gray" surface (i.e., the surface of a graybody).

2.11 Show that Eq. (2.45) becomes Eq. (2.46) for the special case of irradiation by a diffuse gray source.

2.12 Show that for a diffuse gray surface, the hemispherical total absorptivity is equal to the hemispherical total emissivity.

2.13 When can we write $\alpha(\lambda, T_A) = \frac{1}{\pi}\int_{2\pi}\alpha(\lambda, T_A, \vartheta_i, \varphi_i)\cos\vartheta_i\, d\Omega_i$?

2.14 The hemispherical spectral absorptivity of a certain surface can be approximated as a Gaussian distribution with wavelength λ (μm) of the form

$$\alpha(\lambda) = \frac{1}{\sqrt{2\pi\sigma^2}}e^{-(\lambda-\mu)^2/2\sigma^2}, \tag{2.103}$$

where the standard deviation σ is 0.2 μm and the mean of the distribution is $\mu = 0.5$ μm. Plot the hemispherical total absorptivity $\alpha(T_S)$ of the surface as a function of the source temperature T_S when it is irradiated by a blackbody source whose temperature ranges from 100 K to 10 000 K.

2.15 The directional total absorptivity of a certain surface in a specified wavelength interval can be approximated as $\alpha(\vartheta_i) = \cos^n\vartheta_i$, where ϑ_i is the incident zenith angle. Plot the hemispherical total absorptivity α of this surface as a function of n, where $0 \leq n \leq 5$, when it is irradiated with light in the specified wavelength interval whose incident intensity is given by $i(\vartheta_i) = 200 \cos\vartheta_i$ W m^{-2}·sr.

2.16 Show that for a diffuse reflector $\rho(\lambda, T_A, \vartheta_i, \varphi_i) = \pi\, \rho(\lambda, T_A, \vartheta_i, \varphi_i, \vartheta_r, \varphi_r)$.

2.17 Show that for diffuse illumination of a surface element dA

$$\rho(\lambda, T_A, \vartheta_r, \varphi_r) = \frac{i_{\lambda,r}(\lambda, T_A, \vartheta_r, \varphi_r)}{i_{\lambda,i}(\lambda)} \quad \text{(diffuse illumination)}. \tag{2.104}$$

2.18 Demonstrate by direct substitution that Eq. (2.68) satisfies Maxwell's equations, Eqs. (2.62)–(2.65), if Eq. (2.69) is valid.

2.19 Verify Eq. (2.82).

2.20 Derive Eq. (2.85) from (2.84) and the Snell–Descartes law, Eq. (2.76).

2.21 Derive Eq. (2.86).

2.22 Provide a formal argument that the bidirectional spectral reflectivity as defined by Eq. (2.47) cannot be applied to specular surfaces.

2.23 Derive Eq. (2.89) from Eqs. (2.85)–(2.88).

2.24 Eliminate ϑ_t from Eq. (2.89) in favor of n_1/n_2 using the Snell–Descartes law, Eq. (2.76).

2.25 Use the result from Problem 2.24 with Eq. (2.90) to reproduce the curve in Figure 2.20.

2.26 Use the results from Problem 2.24 to estimate by numerical integration the hemispherical spectral emissivity of black glass at a wavelength of 0.546 μm.

2.27 Use Eq. (2.93) to reproduce the curve in Figure 2.21.

2.28 Use Eq. (2.93) to estimate by numerical integration the hemispherical spectral emissivity of pure platinum at a wavelength of 2.0 μm (n = 5.29, k = 6.71).

References

1. Mahan, J.R. (2002). *Radiation Heat Transfer: A Statistical Approach*. New York: Wiley.
2. Stefan, J. (1879). Über die Beziehung zwischen der Wärmstralung und der Temperatur. *Sitzungberichte, Akademiede Wissenschaften* 79 (2): 391–428.
3. Boltzmann, L. (1884). Ableitung des Stefan'schen Gesetzes, betreffend die Abhängigkeit der Wärmstralung von der Temperaturaus der electromagnetischen Lichttheorie. *Annalen der Physik und Chemie* 22: 291–294.
4. Wien, W. (1896). Über die Energievertheilung im Emissionspektrum eines schwarzen Körpers. *Annalen der Physik* 58: 662–669.
5. Rayleigh, Lord (1900). Remarks upon the law of complete radiation. *Philosophy Magazine* 49: 539–540.
6. Jeans, J.H. (1905). On the partition of energy between matter and æther. *The London, Edinburgh, and Dublin Philosophical Magazine and Journal of Science* 10 (55): 91–98.
7. Planck, M. (1901). Über das Gesetz der Energievertheilung im Normalspektrum. *Annalen der Physik*, Band 4: 553–563.
8. Feynman, R.P. (2006). *QED: The Strange Theory of Light and Matter*. Princeton and Oxford: Princeton University Press.
9. Gallimore, S.D. (May 2001). *A Study of Plasma Ignition Enhancement for Aeroramp Injectors in Supersonic Combustion Applications*, PhD Dissertation. Blacksburg: Department of Mechanical Engineering, Virginia Tech.
10. Maxwell, J.C. (1865). A dynamical theory of the electromagnetic field. *Philosophical Transactions of the Royal Society of London* 492–512.
11. Brandenburg, W.M. (1963). The reflectivity of solids at grazing angles. In: *Measurement of Thermal Radiation Properties of Solids* (ed. J.C. Richmond). NASA SP-31.

12. Lide, D.R. (1998). *CRC Handbook of Chemistry and Physics*, 79e. Cleveland: CRC Press LLC Weaver, J. H., and H. P. R. Frederikse, "Optical properties of metals and semiconductors" (Figure 4.7), © CRC Press, with permission.
13. Dulong, P.-L. and Petit, A.-T. (1817). Des recherches sur la mesure des températures et sur les lois de la communication de la chaleur (Seconde partie, des lois du refroidissement). *Anales de Chimie et de Physique* 7: 247.

3

The Radiation Distribution Factor for Diffuse-Specular Gray Surfaces

This chapter introduces the concept of the radiation distribution factor, defined as the fraction of power emitted from one surface of an enclosure that is absorbed by another surface due to direct radiation and to all possible reflections. The distribution factor can be thought of as the sensitivity of the heat absorbed by a given surface to the heat emitted from any other surface. Thus, it can be used not only to compute the distribution of heat throughout an enclosure due to radiation, but also as a design tool targeting selected surfaces to determine their relative exposure to radiative heating. We consider enclosures composed of diffuse-specular gray surfaces; firstly, because of their pedagogical importance and also because of their relevance in preliminary design.

3.1 The Monte Carlo Ray-Trace (MCRT) Method and the Radiation Distribution Factor

The Monte Carlo ray-trace (MCRT) method is a two-step process. The first step involves estimation of the radiation distribution factor matrix

---

The Monte Carlo Ray-Trace Method in Radiation Heat Transfer and Applied Optics,
First Edition. J. Robert Mahan.

This Work is a co-publication between John Wiley & Sons Ltd and ASME Press.
Companion website: www.wiley.com/go/mahan/monte-carlo

D_{ij}, and the second step involves multiplication of D_{ij} by a vector whose components are the source strengths of the surfaces making up an enclosure. Individual elements of the distribution factor matrix may be interpreted as the sensitivity of the power absorbed by surface j to the power emitted by surface i; that is,

$$D_{ij} \equiv \partial Q_{ij} / \partial Q_i, \tag{3.1}$$

where Q_{ij} is the total power in watts emitted from surface i that is absorbed on surface j, and Q_i is the total power emitted from surface i. If Q_i, the total power emitted from surface i, is known and the distribution factor matrix D_{ij} is available for any combination of two surfaces i and j, then the heat absorbed by surface j is

$$Q_j = \sum_{i=1}^{n} Q_{ij}, \quad 1 \leq j \leq n, \tag{3.2}$$

where n is the total number of surfaces and

$$Q_{ij} = Q_i D_{ij}. \tag{3.3}$$

Calculation of Q_i for a given surface condition has already been treated in Chapter 2. The current chapter deals with calculation of the distribution factor D_{ij} and its subsequent use in determining the distribution of thermal radiation among surfaces of an enclosure.

If we can somehow obtain the D_{ij} matrix, we already have the answer to one of the most pressing questions in optical and thermal design: "How sensitive is the heat absorbed by a specific surface j to the heat emitted from a specific surface i?" Consideration of Eqs. (3.2) and (3.3) reveals that calculation of the power distribution among the surfaces of an enclosure is a straightforward vector multiplication once the distribution factor matrix is known. Knowledge of the distribution factor matrix greatly facilitates thermal or optical design because it permits targeted analysis of heat transfer among a limited number of surfaces of particular interest.

3.2 Properties of the Total Radiation Distribution Factor

It can be demonstrated (see Problems 3.1–3.3) that, subject to the gray-body assumption defined in Chapter 2, the total radiation distribution factor has the following useful properties:

1. Conservation of energy:

$$\sum_{j=1}^{n} D_{ij} = 1, \quad 1 \le i \le n \tag{3.4}$$

2. Reciprocity:

$$\varepsilon_i A_i D_{ij} = \varepsilon_j A_j D_{ji}, \quad 1 \le i \le n, \quad 1 \le j \le n \tag{3.5}$$

3. Combination of 1 and 2:

$$\sum_{i=1}^{n} \varepsilon_i A_i D_{ij} = \varepsilon_j A_j, \quad 1 \le j \le n \tag{3.6}$$

In Eqs. (3.4)–(3.6), n is the number of surface elements making up the enclosure, ε is the emissivity, and A is the surface area.

Equation (3.6), which is obtained by summing both sides of Eq. (3.5) over the index i and then substituting Eq. (3.4) into the result, is useful for detecting and eliminating errors made during calculation of the distribution factors for an enclosure. It can also be used to provide a statistically meaningful measure of the accuracy with which the distribution factor matrix for a given enclosure has been computed. The conservation of energy relationship, Eq. (3.4), and the reciprocity relationship, Eq. (3.5), are also useful for detecting errors or for finding unknown distribution factors from known distribution factors using distribution factor algebra. However, note that these relationships cannot be used both for error detection and for finding unknown distribution factors in the same enclosure.

Finally, we note that distribution factors can also be defined for radiation entering the enclosure through an opening o with a specified directional distribution; e.g., collimated in a specific direction. In this case, the appropriate relation for defining the distribution of radiation on the surface elements making up the enclosure is

$$Q_{oj} = Q_o D_{oj}, \quad 1 \le j \le n, \tag{3.7}$$

where Q_o is the power (W) entering the enclosure through opening o and D_{oj} is the fraction of this power absorbed by surface element j.

3.3 Estimation of the Distribution Factor Matrix Using the MCRT Method

The Monte Carlo ray-trace method is a statistical approach in which the analytical solution of a problem is avoided in favor of a numerical simulation whose outcome is expected to be the same as that of the analytical solution but which is easier to obtain. In practice, this often means that the numerical simulation is obtainable while the equivalent analytical solution is for all practical purposes unobtainable.

In the case of a thermal radiation problem, a given quantity of energy is uniformly divided into a large number N_i of discrete energy bundles. Here, we feel no obligation to distinguish between energy and power, as in the steady state the former is the latter multiplied by an appropriate time interval. The N_i energy bundles are followed from their emission by surface element i (or from their entry into the enclosure through opening o), through a series of reflections from other surface elements, to their absorption on surface element j of the enclosure. The optical models of the enclosure walls and the laws of probability are used to determine the number of energy bundles N_{ij} absorbed by a given surface element j, where $j = i$ is a possibility.

A consequence of the definition of the radiation distribution factor is that its value approaches the ratio of N_{ij} to N_i as N_i increases; that is,

$$D_{ij} \approx N_{ij}/N_i. \tag{3.8}$$

The accuracy with which N_{ij}/N_i estimates D_{ij} depends on the number of energy bundles traced from surface i. Of course, as in any model-based analysis, it also depends on the accuracy with which the enclosure geometry and the surface optical models are known. Furthermore, as shown in Chapter 7, the uncertainty associated with the estimate corresponding to the value of N_i can be stated to within a specified confidence interval. Moreover, the ultimate accuracy of the solution to a radiation heat transfer problem using the MCRT method depends in a statistically meaningful way on the product of the number of surfaces n making up the enclosure and the number of rays N traced per surface. The ability of the MCRT method to attribute a confidence level to the uncertainty of the results obtained is a compelling argument for its use.

The development presented in the current chapter is valid only for the case of diffuse-specular gray enclosures. Extension to enclosures whose surfaces are directional emitters and absorbers and bidirectional reflectors, and whose emissivity, reflectivity, and absorptivity vary with wavelength is the topic of Chapter 4. The redistribution of thermal radiation by passage through optically participating media such as cloudy atmospheres, jet engine and rocket plumes, and optical elements constructed of glass is treated in Chapter 5.

We defined the concept of a "surface" in Section 1.4, but now it is convenient to narrow the original definition to incorporate the following two restrictions:

1. The equation for the curve connecting any two points on the surface is a continuous, real, once-differentiable function.
2. The mechanical and chemical treatment of the surface is uniform; that is, the coefficients of the emission, reflection, and absorption models do not vary with position on the surface.

Neither of these restrictions actually intervenes when calculating the elements of the distribution factor matrix; rather, they provide guidance for locating seams between surfaces.

The path followed by an energy bundle from its emission, through its subsequent reflections, until its eventual absorption, is familiarly referred to as a "ray," which explains the use of the terminology "ray-trace" analysis. During a ray trace, whenever it is determined that the path of an energy bundle intersects a surface of the enclosure, the ultimate fate of the energy bundle is determined by treating the surface model coefficients as probabilities. One of two equivalent procedures may be followed in ray tracing.

1. Following the first procedure, the energy bundle retains its original energy from one reflection to the next, with all of its energy ultimately being absorbed by a single surface. At each reflection, a random number is drawn whose value is uniformly distributed between zero and unity. If the value of the random number exceeds the local absorption coefficient, the energy bundle is deemed to be reflected; otherwise it is absorbed.
2. Following the second procedure, a fraction of the energy carried by the energy bundle is left behind at each reflection, and the ray trace is terminated when a predetermined fraction of the original energy remains.

Both of these approaches to ray tracing have their particular advantages and both are commonly used.

It is instructive to begin by formulating the first, or so-called "sudden death," method, in which all of the energy carried by the bundle is absorbed during a single absorption event, typically after a number of reflections. We consider an enclosure consisting of n surfaces and we seek the elements of the distribution factor matrix D_{ij}, where

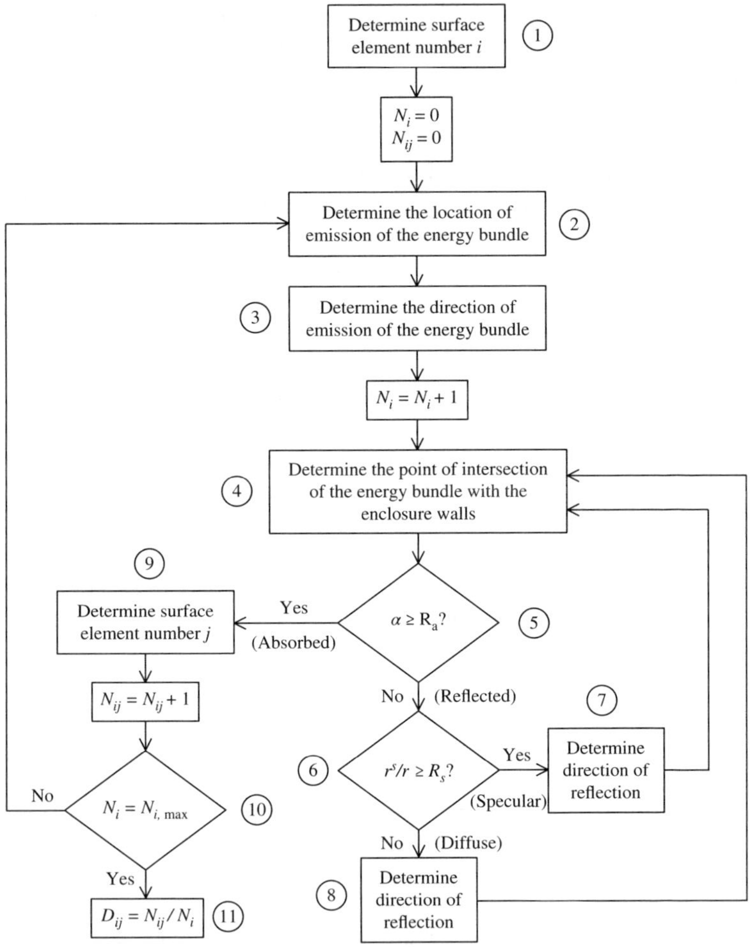

Figure 3.1 Logic block diagram of the Monte Carlo ray-trace method for diffuse-specular gray enclosures.

$1 \leq i \leq n$ and $1 \leq j \leq n$. The logic block diagram corresponding to this formulation appears in Figure 3.1. Modification of this block diagram to accommodate the second method, commonly referred to as the "slow death" method, is left as an exercise for the reader (see Problem 3.4).

The steps enumerated in this section, which correspond to the step numbers in Figure 3.1, are established in the context of the optical and thermal-radiative design of the instrument concept appearing in

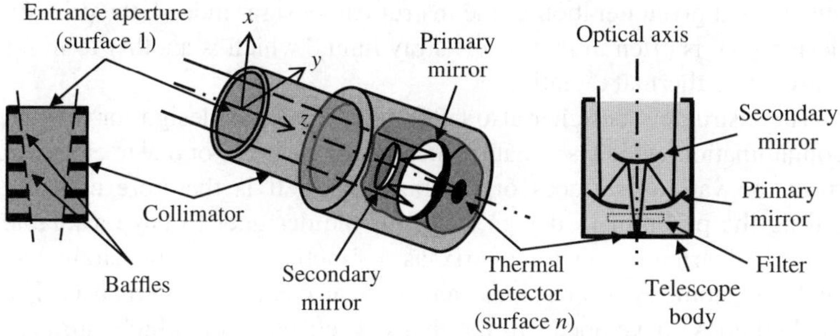

Figure 3.2 A concept for remote measurement of thermal radiation.

Figure 3.2. The sensitivity of the radiant heat flux absorbed by a given surface to the flux emitted by each of the other surfaces of the enclosure is sought. In the current example, surface n is a thermal radiation detector of some sort, while the remaining surfaces of the enclosure constitute a collimator attached to a Cassegrain telescope intended to focus light on the detector. The entrance aperture of the instrument shown in the figure is the imaginary disk-shaped surface located at the outer end of the collimator (Surface 1). Definition of the distribution factor D_{1n} is of obvious interest to the instrument designer, as it represents the transfer function relating the heat flux incident to the instrument aperture to the heat flux absorbed by the detector. That is, it is the instrument optical gain.

Most of the thermal radiation entering through the aperture is intercepted by the tapering system of ring-shaped baffles that make up the collimator and define the instrument field-of-view. The idea is to limit radiation reaching the detector to angles that deviate only slightly – a couple of degrees or less – from the optical axis. The surfaces within the collimator are as black as can be rendered by commercially available coatings, ensuring that they absorb upwards of 90% of the radiation incident to them. However, some of the radiation initially incident to the baffles and to the cylindrical walls of the collimator – typically on the order of 10% – is reflected, and some of this scattered radiation – perhaps as much as 0.1% depending on the design of the telescope – inevitably finds its way to the detector. This means that a small but measureable fraction of the radiation reaching the detector comes from "out of field"; that is, from angles with respect to the instrument optical axis beyond those defined by a "perfect" collimator. It is of interest to the instrument designer to be able to predict this out-of-field component of the detector signal for a given iteration of the instrument design. Indeed, the primary design goal is often to minimize "stray light," which is another term for out-of-field thermal radiation.

The instrument designer also needs to evaluate the design for thermal contamination, which is radiation incident to the detector due to emission from the various surfaces of the instrument. It is therefore essential during the preliminary design stage for the designer to have available the full distribution factor matrix as a function of design parameters such as geometry, dimensions, and optical coating performance. The methodology developed in the current chapter, in which surfaces are assumed to be diffuse-specular and gray, is usually adequate for preliminary design. Extensions to the more realistic case of bidirectional wavelength-dependent surface behavior – considered in Chapter 4 – may be required to verify that the final design meets specifications. This is especially true if the diffuse-specular gray analysis indicates marginal performance.

***Step 1*: Determine the index number of the surface element from which the energy bundle is to be emitted.**

For the stray-light analysis described above, we are primarily interested in determining the distribution factor D_{1n}. In a practical design the ratio of the area of the entrance aperture to that of the detector would be large – of the order of 100 : 1. Also, both surfaces would be essentially black[1], meaning that for all practical purposes $\alpha = \varepsilon = 1.0$. Therefore, the reciprocity relation, Eq. (3.5), can be written

$$\varepsilon_1 A_1 D_{1n} = \varepsilon_n A_n D_{n1}, \tag{3.9}$$

yielding

$$D_{1n}/D_{n1} = A_n/A_1 << 1.0. \tag{3.10}$$

In other words, D_{n1} may be anticipated to be much larger than D_{1n}. Referring to Eq. (3.8), if a fixed number of rays is to be traced from each surface element – that is, if $N_1 = N_n$ – it follows that N_{n1} will be much larger than N_{1n}. We conclude that it is a better use of resources to compute D_{n1} rather than D_{1n}, because the error in the result will be inherently much less in the former case than in the latter. To understand this assertion, consider that if only 10 rays entering through the aperture

[1] *All* radiation incident to the entrance aperture from inside the instrument will escape, while the detector surface will be as black as current technology allows. Depending on wavelength, a state-of-the-art detector coating can be expected to absorb as much as 98% of the radiation incident to it.

are absorbed by the detector, while perhaps 1000 rays emitted by the detector escape through the entrance aperture – as would be approximately the case if $A_1/A_n = 100$ – then a difference of one ray translates to a 10% uncertainty in determining D_{n1} but only a 0.1% uncertainty in determining D_{1n}. Therefore, we trace rays from the detector to the entrance aperture to determine D_{n1} and then use reciprocity, Eq. (3.5), to compute D_{1n}; that is,

$$D_{1n} = (A_n/A_1)\,D_{n1}. \tag{3.11}$$

In the case at hand we seek only the elements of the distribution factor matrix D_{in}, where $1 \leq i \leq n$ with Surface 1 being the entrance aperture. The detector typically would have the smallest area among all of the surfaces comprising the instrument. Therefore, following the above logic, the surface number i chosen in this step is n. That is, we trace a large number of rays – one million is typical – from the detector to each of the surfaces of the instrument enclosure, including the imaginary Surface 1. Alternatively, if a full radiation heat transfer analysis among all surfaces of the enclosure is to be carried out, it would be necessary to cycle n times through Step 1, allowing i to vary from 1 to n, where Surface 1 is the entrance aperture.

Finally, before moving on to Step 2 it is necessary to initialize counters for N_i and N_{ij} to zero.

***Step 2*: Determine the coordinates *x*, *y*, and *z* of the emission point of the energy bundle.**

The details in this step vary depending on the shape of the emitting surface. The emitting surface in the example under consideration is the detector element, which is a disk. In order to determine the point of emission from a surface, it is first necessary to establish a global coordinate system. The global coordinate system has two essential properties:

1. It is always specified in terms of right-handed Cartesian coordinates. Ray-trace analysis is a vector operation and digital computers in general, and Matlab in particular, are most efficient when working with matrices and vectors.
2. Once established, all surface coordinates and ray directions are defined relative to it.

In the case at hand, it is convenient to anchor the global coordinate system in the center of the entrance aperture with the $+z$-axis directed into the instrument, as indicated in Figure 3.2. The components of all vectors are aligned with the x-, y- and z-axes, respectively.

For emission from a uniform disk whose diameter is d and whose center is located at (x_c, y_c, z_c), we seek the randomly positioned source point

$$x_0 = x_c + r\cos\varphi, \quad y_0 = y_c + r\sin\varphi, \quad \text{and} \quad z_0 = z_c; \tag{3.12}$$

where $r = (d/2)\sqrt{R_r}$ and $\varphi = 2\pi R_\varphi$.

In Eq. (3.12), R_r and R_φ are two successively drawn random numbers uniformly distributed between zero and unity, and the azimuth angle φ is measured in the anticlockwise sense with respect to the x-axis. When applied to the current instrument numerical model in which $(x_c, y_c, z_c) = (0, 0, z_{fp})$, where z_{fp} is the z-coordinate of the instrument focal plane, Eq. (3.12) becomes

$$x_0 = r\cos\varphi, \quad y_0 = r\sin\varphi, \quad \text{and} \quad z_0 = z_{fp}. \tag{3.13}$$

***Step 3*: Determine the direction cosines *L*, *M*, and *N* of the energy bundle.**

The concept of direction cosines has already been introduced and explained in Chapter 1. Once again, the details in this step depend on the shape of the emitting surface. For a plane, a diffuse surface such as the disk-shaped detector, whose center is at $(x, y, z) = (0, 0, z_{fp})$ and whose unit normal vector is $-\boldsymbol{k}$, the procedure is relatively uncomplicated. We simply set $\boldsymbol{n} = -\boldsymbol{k}$, $\boldsymbol{t}_1 = \boldsymbol{i}$, and $\boldsymbol{t}_2 = \boldsymbol{j}$, after which Eq. (1.34) yields

$$L = \sin\vartheta_E \cos\varphi_E, \quad M = \sin\vartheta_E \sin\varphi_E, \quad \text{and} \quad N = \cos\vartheta_E. \tag{3.14}$$

In Eq. (3.14) the random azimuth angle φ_E is defined

$$\varphi_E = 2\pi R_\varphi, \tag{3.15}$$

where R_φ is once again a uniformly distributed random number; and the random zenith angle θ_E is given by

$$\vartheta_E = \sin^{-1}\left[\sin\vartheta_{\max}\sqrt{R_r}\right], \tag{3.16}$$

where $\vartheta_{\max}$ is the maximum allowed angular deviation with respect to the optical axis of a diffusely emitted ray. Derivation of Eq. (3.16) is deferred until Chapter 4. Limiting the angular range of the radiation entering the instrument aperture is sometimes useful in parametric studies aimed at determining the impact of the source structure on instrument performance. Equation (3.14) is also valid for the case of collimated sources, except that now we designate fixed values of ϑ_E and φ_E rather than values computed using Eqs. (3.15) and (3.16).

The generic process that must be followed in the case of more complicated surface geometries is addressed in Step 8 in the context of reflections.

Once Step 3 has been executed, the ray counter N_i is incremented before moving on to Step 4.

***Step 4*: Determine the point of intersection of the current ray with the enclosure walls.**

This step is the heart and soul of the MCRT method, occupying most of the processor time required to execute a ray trace. The processor time requirement is directly proportional to the number of surface elements making up the enclosure. This is because a given ray, once launched, must visit every surface of the enclosure looking for a valid intersection. Why? Because all surface elements are created by "trimming" a member of a relatively small class of master surfaces, and each master surface is in principle infinite in extent. This means that every ray could, in principle, intersect every master surface figuring in the design of the enclosure. More to the point, no dependable rule exists to justify *a priori* elimination of a given surface from the search.

The details for a given enclosure are geometry-specific but the generic approach is always the same:

a. Sequentially visit each surface element making up the enclosure and search the element for points of intersection with the ray. This requires identifying the master surface type (plane, cylinder, sphere, cone, paraboloid, hyperboloid, ellipsoid, etc.) for each surface and then solving the corresponding surface equation simultaneously with the equations for the line containing the ray, as described in Section 1.3. The results of this process are candidate values for the coordinates (x_1, y_1, z_1) of the point(s) of intersection and their distance(s) t_1 from the source point (x_0, y_0, z_0).
b. Verify whether or not the point(s) of intersection lies within the known range of the master surface from which the surface element is trimmed.
c. Eliminate any points of intersection that are on the "wrong" side of the surface.
d. Eliminate any back candidates that may exist.
e. Compare the current value of t_{new} with the smallest value t_{old} obtained so far during visits to previous surfaces. Set $t_1 = t_{new}$ if $t_{new} < t_{old}$; otherwise, $t_1 = t_{old}$. Update the values of (x_1, y_1, z_1) to the values corresponding to t_1 and continue to the next surface.
f. Continue to Step 5 when all n surfaces have been visited.

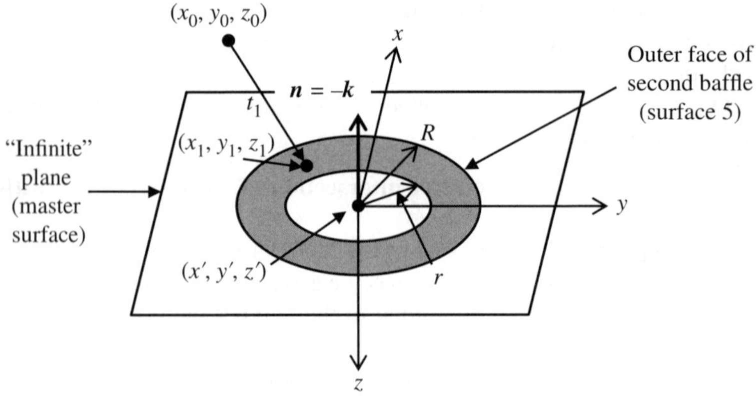

Figure 3.3 Outer face of the second baffle of the collimator in Figure 3.2.

As an example of this process, suppose Surface 5, shown in Figure 3.3, is the washer-shaped surface on the face of the second baffle oriented toward the entrance aperture in the collimator. Then in Step 4a we seek the intersection of the line

$$(x_1 - x_0)/L_0 = (y_1 - y_0)/M_0 = (z_1 - z_0)/N_0 = t_1, \tag{3.17}$$

with the x,y-plane at $z = z_1$ whose surface normal is

$$\boldsymbol{n} = n_x \boldsymbol{i} + n_y \boldsymbol{j} + n_z \boldsymbol{k} = -\boldsymbol{k}. \tag{3.18}$$

It is shown in Chapter 1 that

$$t_1 = [n_x (x' - x_0) + n_y (y' - y_0) + n_z (z' - z_0)]/(n_x L_0 + n_y M_0 + n_z N_0). \tag{3.19}$$

Equation (3.19) is generic; that is, it applies to any plane surface. The point (x', y', z') can be any point in the master plane, even if it lies outside the range of the surface element in question. In the case at hand it is convenient to set $(x', y', z') = (0, 0, z_w)$, where z_w is the known z-coordinate of the "washer" whose inner radius is r and whose outer radius is R. Equation (3.19) then simplifies to

$$t_1 = (z' - z_0)/N_0, \text{ where } z' = z_1 = z_w. \tag{3.20}$$

This result is a reminder that the plane is the only surface type guaranteed to yield a single non-degenerate intersection with a straight line.

Equation 1.5 may then be written

$$x_1 = x_0 + L_0\,t_1, \quad y_1 = y_0 + M_0\,t_1, \quad \text{and} \quad z_1 = z_0 + N_0\,t_1 = z_w. \tag{3.21}$$

Recall that all second-order surfaces will generally return two candidate points of intersection. The fact that the plane surface of our current example returns only one candidate greatly simplifies the following discussion.

In Step 4b we determine whether or not point (x_1, y_1, z_1) lies on the washer-shaped surface element. We accomplish this by asking if $r^2 \leq {x_1}^2 + {y_1}^2 \leq \mathrm{R}^2$. If TRUE we move on to Step 4c. However, if FALSE we set t_1 to an arbitrary value sufficiently large to ensure that it will exceed any possible legitimate value. That is, we must be sure that it will not survive the test administered in Step 4e. In the FALSE case we can skip Steps 4c and 4d and move on to Step 4e.

In the special case of a plane surface, Steps 4c and 4d can be combined. This is because if the source point is on the "back" side of a plane surface it cannot illuminate the surface regardless of the direction of the emitted ray, and so the back candidate question is irrelevant. However, when second-order surfaces are involved there will generally be two candidates, in which case Step 4c is a test of whether a given candidate source point is "inside" or "outside" a surface. In the current example we need only ask if $z_0 < z_1$. It is clear that if this proposition is FALSE, the source point must lie on the wrong side of the master surface to illuminate it. In general, if a ray passes the test imposed in Step 4c we move on to Step 4d, but if it fails the test we once again set t_1 to an arbitrary value sufficiently large to ensure that it will exceed any possible legitimate value and move on to Step 4e.

In Step 4d we ask if $\boldsymbol{v} \cdot \boldsymbol{n} \leq 0$, where $\boldsymbol{v} = L_0\,\boldsymbol{i} + M_0\,\boldsymbol{j} + N_0\,\boldsymbol{k}$ and $\boldsymbol{n}$ is the surface unit normal vector given by Eq. (3.18). Recall that the scalar product of any two unit vectors is the cosine of the angle between them. Therefore, if this product is negative, the angle between the two vectors is less than 90°; if the product is zero, the angle between them is 90°; and if the product is positive, the angle between them is greater than 90°. It should be clear that a ray whose direction cosines are (L_0, M_0, N_0) will strike a surface whose unit normal vector components are (n_x, n_y, n_z) if, and only if, the proposition $\boldsymbol{v} \cdot \boldsymbol{n} \leq 0$ is TRUE. In this case we move on to Step 4e. However, if the proposition is FALSE we set t_1 to an arbitrarily large value before moving on to Step 4e.

Only one of all the points of intersection that survive the filtering process imposed by the previous steps will be the shortest distance from the source. This will be the actual point of intersection of the ray with the walls of the enclosure. Step 4e is the sorting process that identifies this point, after which Step 4f transfers control to Step 5.

***Step 5*: Determine whether the ray is absorbed or reflected.**

The next available random number R_α is drawn and its value compared to the value of the absorptivity α of the surface containing point (x_1, y_1, z_1). The justification for adding two restrictions to the definition of a surface now becomes clear: each numbered surface is defined by its own unique set of optical models. If $\alpha \geq R_\alpha$ the ray is absorbed; otherwise, it is reflected.

We have already seen ample justification for use of the term "Monte Carlo" when describing the method elaborated in this book: the location of the point of emission on a surface element and the direction of emission, at least for the case of uniform, diffuse emission, both depend almost literally on the "roll of the dice." Now we see that the laws of chance enter into the process at yet another level: the decision of whether a ray is absorbed or reflected is based on interpretation of the absorption coefficient as the probability that a ray incident to a surface will be absorbed.

If it is determined that the ray is reflected because $\alpha < R_\alpha$, then control is transferred to Step 6; otherwise, it is transferred to Step 9. For the moment we assume that the ray is reflected.

***Step 6*: Determine whether the reflection is diffuse or specular.**

A more sophisticated version of this step is used in the more realistic situation considered in Chapter 4, where surfaces are wavelength-dependent, bidirectional reflectors. The diffuse-specular gray model considered here is adequate for upwards of 90% of practical engineering applications, such as preliminary design of optical and thermal radiative systems. However, greater accuracy is often required for the more accurate simulations considered in Chapter 4. In terms of code development, upgrading from a current diffuse gray model to a bidirectional spectral model is in principle a matter of replacing the *function* (using the nomenclature of Matlab), or subroutine, that implements this step.

At this point it is convenient to introduce the concept of the specularity ratio, ρ^s/ρ, where ρ^s is the specular component of the reflectivity and $\rho = \rho^s + \rho^d$ is the sum of the specular and diffuse components of the reflection coefficient. The idea behind the specularity ratio is well captured in Figure 1.9. The next available random number R_s is drawn and its value compared to the specularity ratio. If $\rho^s/\rho \geq R_s$, the reflection is specular; otherwise, it is diffuse. In the case of a specular ("mirror-like") reflection, control is transferred to Step 7; otherwise, it is transferred to Step 8.

***Step 7*: Determine the direction of the specular reflection.**

The case of specular reflection is fully developed in Section 1.4 and leads to

$$\boldsymbol{v}_R = \boldsymbol{v}_I - 2(\boldsymbol{v}_I \cdot \boldsymbol{n})\,\boldsymbol{n} = \boldsymbol{v}_I + 2\cos(\vartheta)\,\boldsymbol{n}. \tag{1.27}$$

The nomenclature of this equation appears in Figure 1.7. In the current example, the unit normal vector $\boldsymbol{n}$ is equal to $-\mathbf{k}$. Therefore, the direction cosines for the reflected ray are $L_1 = L_0$, $M_1 = M_0$, and $N_1 = (1 - 2\cos\vartheta)$ N_0. Control is now transferred back to Step 4.

***Step 8*: Determine the direction of the diffuse reflection.**

In principle, a diffuse reflection can be treated almost exactly the same way as a diffuse emission, with the only difference being that the emission direction cosines are directly specified in terms of the global (x, y, z) coordinate system, while the reflection direction is first determined in a local ($\boldsymbol{n}$, $\boldsymbol{t}_1$, $\boldsymbol{t}_2$) coordinate system and then translated into global coordinates, as described in Section 1.4. The preferred approach, however, is Kowsary's **tangent sphere method** [1], which offers two significant advantages: (i) it avoids the possibility of dividing by zero when using Eq. 1.29 to compute $\boldsymbol{t}_1$, and (ii) it speeds up the overall enclosure analysis by a factor of about ten. Kowsary's method is illustrated in Figure 3.4,

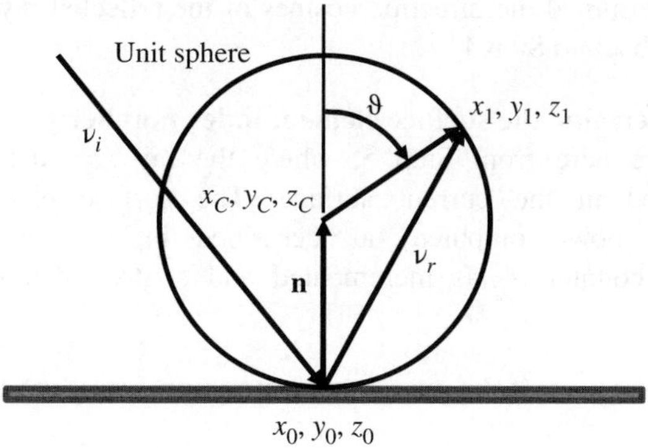

Figure 3.4 Kowsary's method [1] for computing the direction of a diffusely reflected ray.

which is a two-dimensional view of a three-dimensional diffuse reflection.

The point of incidence (x_0, y_0, z_0) and the local unit normal vector $\boldsymbol{n} = n_x\boldsymbol{i} + n_y\boldsymbol{j} + n_z\boldsymbol{k}$ are both known. Then the center of a unit sphere tangent to the surface at (x_0, y_0, z_0) is (x_C, y_C, z_C), where

$$x_C = x_0 + n_x, \quad y_C = y_0 + n_y, \quad \text{and} \quad z_C = z_0 + n_z. \tag{3.22}$$

We then draw the next two available random numbers, R_ϑ and R_φ, and compute

$$\vartheta = \cos^{-1}[2R_\vartheta - 1] \quad \text{and} \quad \varphi = 2\pi R_\varphi. \tag{3.23}$$

Then the unit vector of the diffusely reflected ray is

$$\boldsymbol{v_r} = L_1\boldsymbol{i} + M_1\boldsymbol{j} + N_1\boldsymbol{k} = \frac{(x_1 - x_0)}{t}\boldsymbol{i} + \frac{(y_1 - y_0)}{t}\boldsymbol{j} + \frac{(z_1 - z_0)}{t}\boldsymbol{k}, \tag{3.24}$$

where (x_1, y_1, z_1) is a random point on the unit circle, and t is the distance from (x_0, y_0, z_0) to (x_1, y_1, z_1). It is elementary to show (see Problem 3.6) that

$$(x_1 - x_0) = n_x + \sin\vartheta\cos\varphi, (y_1 - y_0) = n_y + \sin\vartheta\sin\varphi, \quad \text{and}$$
$$(z_1 - z_0) = n_z + \cos\vartheta; \tag{3.25}$$

and

$$t = \sqrt{2(1 + n_x\sin\vartheta\cos\varphi + n_y\sin\vartheta\sin\varphi + n_z\cos\varphi)}. \tag{3.26}$$

Having determined the direction cosines of the reflected ray, control is transferred back to Step 4.

***Step 9*: Determine the surface element index number j.**

We arrive here from Step 5, where the ray was determined to be absorbed in the current surface. The surface element index number is now computed, as described in Section 1.6, after which the counter N_{ij} is incremented and control is transferred to Step 10.

***Step 10*: Determine if the preset maximum number of rays have been traced.**

If the maximum number of rays to be traced from a given surface i has been achieved ($N_i = N_{i,\max}$), we move on to Step 11, where we either compute the estimate of the distribution factor $D_{ij} \approx N_{ij}/N_i$ or "bin" the ray, or both. If $N_i < N_{i,\max}$, we return control to Step 2 and launch the next energy bundle.

***Step 11*: Compute the estimate of the distribution factor, $D_{ij} \approx N_{ij}/N_i$.**

In some cases – and the example under consideration would most likely fall into this category – it is desirable to know not only the value of the estimate of the distribution factor from the emitting surface to the absorbing surface but also the spatial distribution of the absorbed radiation. The spatial distribution is obtained by binning.

3.4 Binning of Rays on a Surface Element; Illustrative Example

Meshing and indexing are treated in Section 1.6. A closely related idea is **binning**, in which indexing allows us to keep track of the spatial distribution of absorbed energy bundles. In the case of the instrument in Figure 3.2, it might be desirable to know how the power incident to the entrance aperture is distributed across the detector face. In this situation, after executing Step 11, control is transferred to a binning step not shown in Figure 3.1. The detector surface, subdivided into a number of bins, is illustrated in Figure 3.5. In a given application the actual number of bins and their spatial distribution would be dictated by the shape of the surface and the desired spatial resolution.

No attempt has been made in creating Figure 3.5 to suggest a particular scheme for determining the spacing between consecutive radial rings. However, two schemes come to mind: (i) either the ring boundaries could be equally spaced or (ii) they could be spaced to achieve equal-area rings. The former scheme is, of course, easy to achieve but is probably

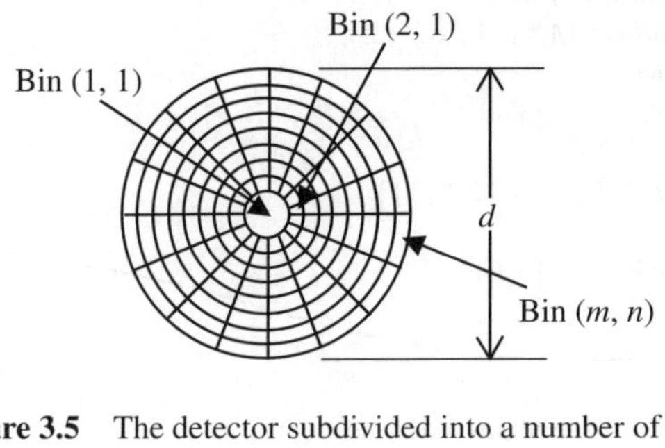

Figure 3.5 The detector subdivided into a number of bins.

not as useful as the latter. A plot of power absorbed as a function of radial position would be easier to interpret if the rings were divided into equal areas. This may be achieved using the formula

$$r_m = (d/2)\sqrt{m/M}, \tag{3.27}$$

where d is the diameter of the detector, M is the total number of radial bands to be created, and r_m is the radial position of the outer radius of the mth band. It is both natural and convenient to base the circumferential divisions on equal angles,

$$\varphi_n = 2\pi n/N, \tag{3.28}$$

where N is the total number of angular sectors, n is the nth such sector, and φ_n is the angle corresponding to the upper angular limit of the sector.

We arrive at the binning step with knowledge of x_1, y_1, z_1, and P_1, where P_1 is the power carried by the current ray. The Matlab *function* for determining the values of m and n and updating the power accumulated in bin (m, n) appears in Figure 3.6. The "floor" operator on lines 8 and 12 of Figure 3.6 returns the least integer in the argument.

```
1  function [ m, n, P ] = BIN( x1, y1, P1, P)
2  r1 = sqrt( x1 ^ 2 + y1 ^ 2) ;
3  if r1 <= ( d / 2 ) / sqrt( M )
4     m = 1;
5     n = 1;
6  else
7     phi1 = atan( y1 / x1 ) ;
8     m = floor( 4 * M * ( r1 / d ) ^ 2 );
9     if m >= M
10       m = M;
11    end
12    n = floor( N * phi1 / ( 2 * pi ) );
13 end
14 P( m, n ) = P( m, n ) + P1;
15 end
```

Figure 3.6 Matlab binning pseudo-code.

3.5 Case Study: Thermal and Optical Analysis of a Radiometric Instrument[2]

Figure 3.7 is a detailed data sheet for the radiometric instrument concept illustrated in Figure 3.2. Our task is to provide a preliminary evaluation of its optical and thermal performance. The approach is to model the

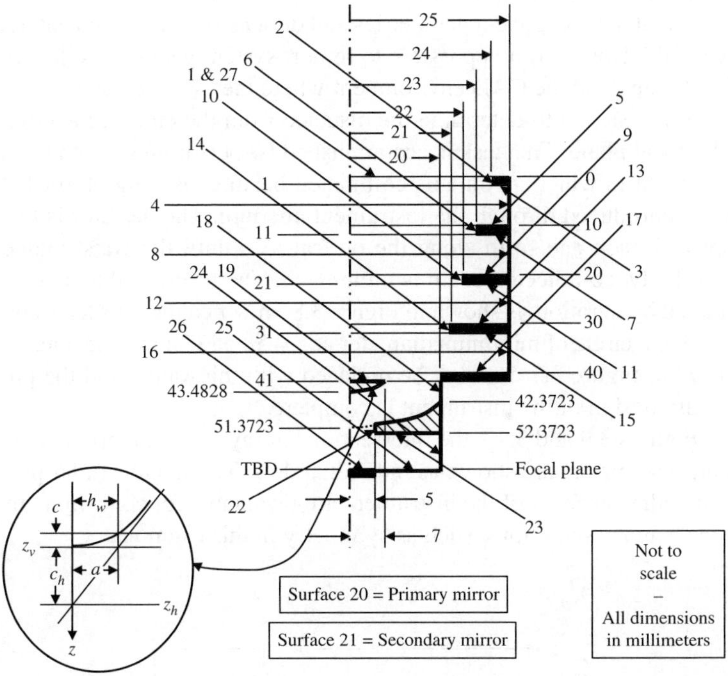

For all surfaces *except* the primary mirror, the secondary mirror, the detector, and the entrance aperture, $a = 0.9$ and $r^s/(1 - a) = 0.9$. The mirrors are ideal specular reflectors, and the detector and entrance aperture are ideal absorbers.

Primary mirror prescription: (elliptic paraboloid)	$S(x, y, z) = x^2 + y^2 - r(z_v - z) = 0$, where $r = a^2/c$; $a = 20$ mm, $c = 9$ mm, vertex at $z_v = 51.3723$ mm.
Secondary mirror prescription: (hyperboloid)	$S(x, y, z) = x^2 + y^2 - (a/c_h)^2(z - z_h)^2 + a^2 = 0$; vertex at $z_v = 43.4828$, $c = 2.4828$ mm, $h_w = a = 7.0$ mm, $c_h = c/\{[1 + (h_w/a)^2]^{1/2} - 1\} = 5.9939$ mm, $z_h = z_v + c_h$

Figure 3.7 Data sheet for radiometric instrument concept of Figure 3.2.

[2] This case study is based on a student term project report by Jonathan Pfab, Virginia Tech, Spring 2017.

optical behavior of the instrument under the assumption of specularly reflecting mirrors and diffuse-specular non-mirror surfaces in the visible wavelength range (0.4–0.7 μm), and to model the thermal behavior of the instrument in the infrared. Optical characterization will require assessment of baffle and telescope performance, while thermal characterization will require estimation of the distribution factor matrix. The listing for the Matlab code developed to accomplish this task is available at the companion website listed on p. xix. Study of the code listing reveals that the component geometries and dimensions have been entered "by hand." However, it is possible to import system geometries directly into Matlab from the CAD environment where they were created.

The first step is to determine the distance from the entrance aperture to the focal plane. This task is accomplished by performing a numerical experiment in which an on-axis collimated beam consisting of 100 000 rays is introduced through the instrument aperture. The detector is then displaced back and forth along the optical axis until the waist diameter of the twice-reflected beam is minimized. The result of this process, obtained by iteration, is shown in Figure 3.8, in which the rays have converged to a circle of minimum diameter at $z = 58.5823$ mm. The notation "TBD" in Figure 3.7 can now be replaced with this value, and the preliminary design of the instrument is complete.

In Figures 3.9 and 3.10, the intersections of rays with the primary and secondary mirrors are shown as individual dots. These and similar plots for the other surfaces of the instrument are invaluable for confirming the basic design assumptions, such as symmetry in this example.

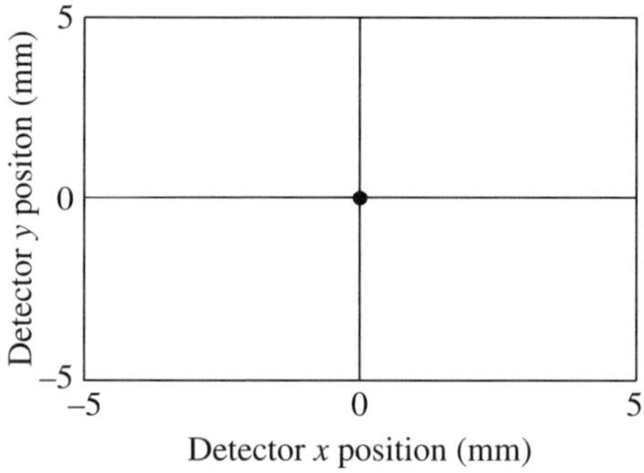

Figure 3.8 One hundred thousand rays converging on the focal plane at $z = 58.58$ mm.

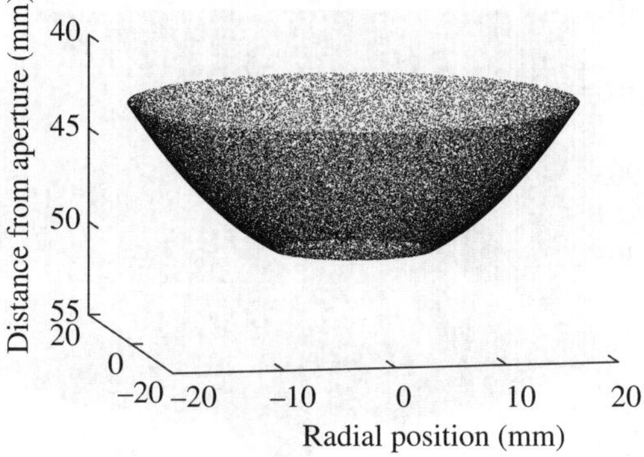

Figure 3.9 Ray intersections with the primary mirror shown as individual dots.

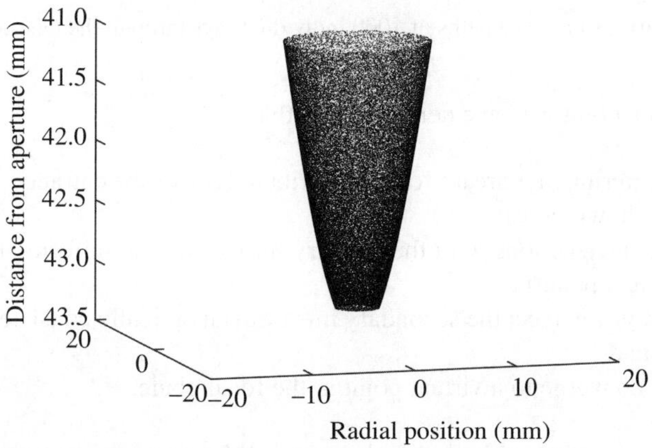

Figure 3.10 Ray intersections with the secondary mirror shown as individual dots.

Another way of confirming the basic design is to trace the paths of individual rays through the telescope, as illustrated in Figure 3.11. Only 1000 rays are traced in this case so that individual rays can be discerned. For clarity, the baffle surfaces have been removed in creating this figure. The focal point of the telescope is clearly revealed in this presentation format. Inspection of the code listing at the companion website confirms that all of the figures in this case study have been created using standard Matlab plotting tools.

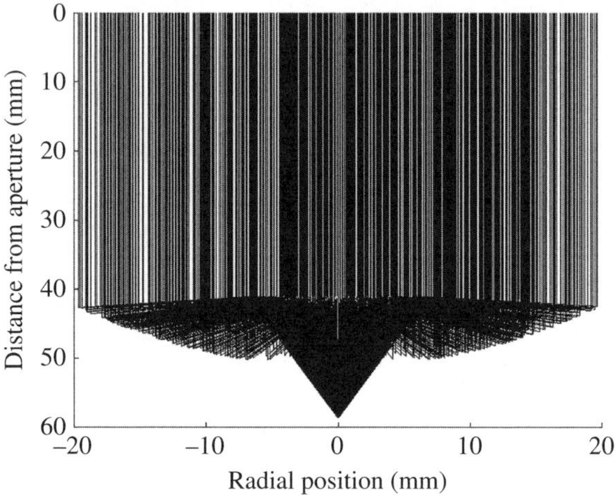

Figure 3.11 The paths of 1000 individual rays through the telescope.

To this point we have demonstrated that:

1. the entering rays are uniformly distributed across the entrance aperture (not shown here),
2. their intersections with the primary mirror form a uniform and symmetrical pattern,
3. the rays intersect the secondary mirror in an optically consistent manner, and
4. they converge to a virtual point in the focal plane.

Having confirmed both the ideal optical design of the instrument and the correct application of the ray-trace logic used so far, we are now ready to enhance the instrument model by adding the collimator and other non-optical surfaces. The modular architecture of a typical MCRT model is one of its most important features. The ability to incrementally create a system model from its components in a step-by-step process allows validation of each successive module before passing on to the next step, or module. Therefore, we add the surfaces making up the collimator one at a time, following the order of the numbering system given in Figure 3.7 and testing the code after the addition of each new surface.

Rather than modeling the source of radiation as a collimated beam, which was convenient for verifying optical performance, we now model it as a diffuse disk located in the telescope entrance aperture (at $z = 0$

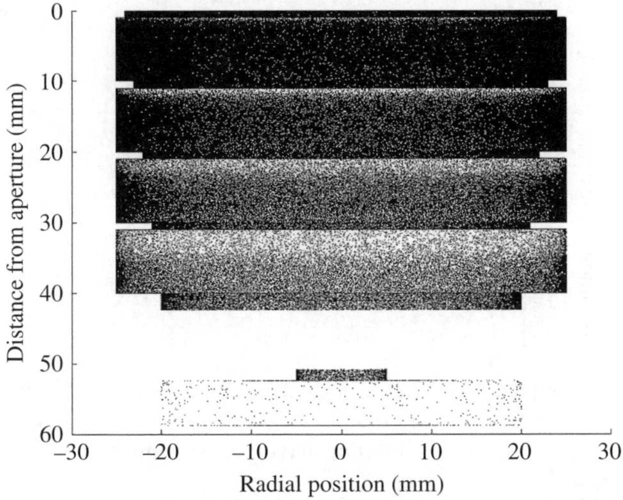

Figure 3.12 Absorption of radiation on the surfaces of the collimator, on the inner edge of the primary mirror (Surface 22 in Figure 3.7), and on the focal plane.

in Figure 3.7). This allows us to evaluate the ability of the collimator to reject (by absorption) "out-of-field" radiation. Figure 3.12 shows the distribution of rays absorbed in the collimator, on the edge of the primary mirror (Surface 22 in Figure 3.7), and on the focal plane. We see that the density of absorbed light generally decreases as we move from top to bottom of the figure, or into the telescope. This is consistent with the diffuse distribution of the source described by Eq. (1.32).

Of the one million diffusely distributed rays entering the collimator, only 8360 are absorbed in the focal plane, and of these only 4828 are incident to the detector. Figure 3.13 shows the spatial distribution of the rays absorbed in the focal plane, with the rays absorbed on the detector shown in black. The significant conclusion to be drawn from this figure is that the spatial distribution on the detector appears to be uniform. This conclusion can be verified by subdividing the detector into equal-area bins, as discussed in Section 3.4.

The rays absorbed in the focal plane after only one reflection from each of the two mirrors and no other reflections are referred to as "optical" rays, while rays that have suffered at least one reflection from a collimator surface or multiple reflections from one of the two mirrors are referred

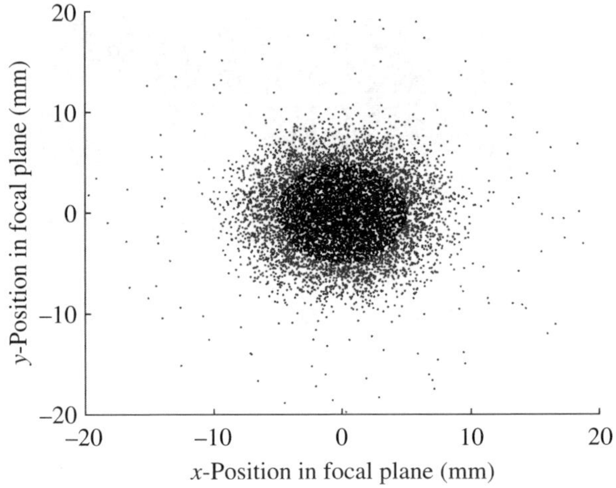

Figure 3.13 Distribution of rays absorbed in the focal plane (gray dots) and on the detector (black dots).

to as "non-optical." Identification and minimization of non-optical rays, or "stray light," incident to the detector is of cardinal interest in radiometric design. Of the 4828 rays absorbed by the detector, 4477 (92.7%) are optical. Figure 3.14 shows the distribution of non-optical rays in the focal plane, with those absorbed by the detector gathered in the center. Comparison of Figures 3.13 and 3.14 suggests that the size of the detector could be increased, thereby increasing the signal without increasing the signal-to-noise ratio.

Figure 3.15 shows the variation of detector power with angle of incidence of a collimated beam, consisting of 100 000 rays, that floods the telescope entrance aperture. From this figure we may conclude that the collimator effectively rejects radiation entering the instrument at angles greater than about 6°. In a typical application it would be of interest to minimize the "out-of-field," or "stray light," contribution to the detector signal. That is, the goal would be to maximize the ratio of the power absorbed for incidence angles less than 6° to the power absorbed for greater incidence angles. The MCRT method would clearly play an essential role in any design optimization scheme aimed at accomplishing this goal.

If the instrument was intended to form an image, the detector might consist of a two-dimensional array of microbolometers (called

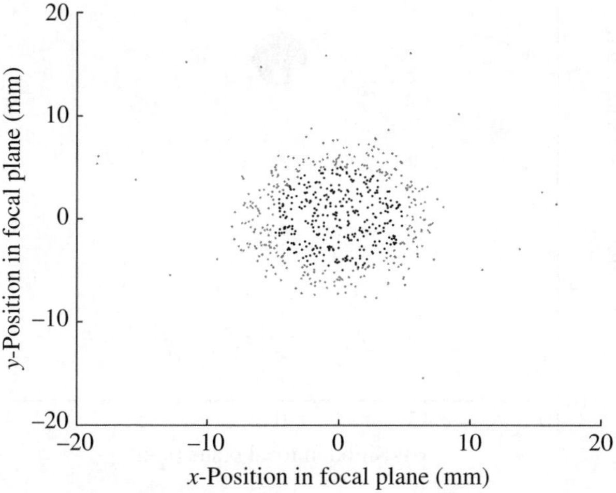

Figure 3.14 Distribution of non-optical rays in the focal plane (rays absorbed by the detector shown in black are gathered in the center).

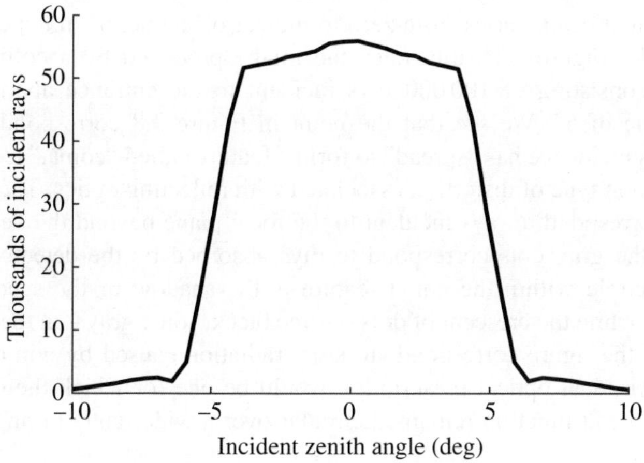

Figure 3.15 Variation of detector power with angle of incidence of a collimated beam.

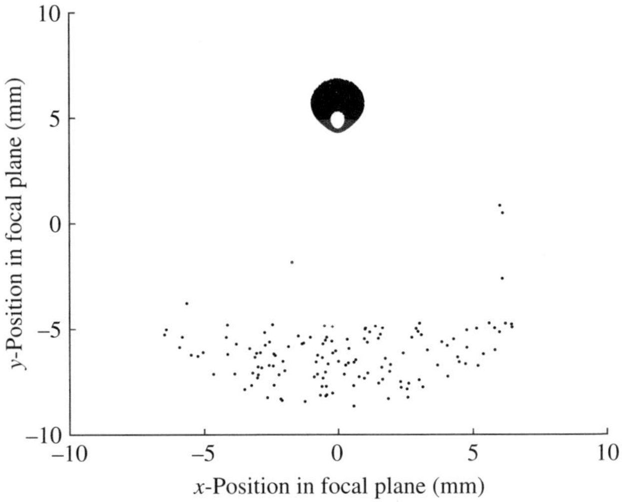

Figure 3.16 Coma produced by a collimated beam of rays incident to the instrument aperture at an angle of 5°.

a "focal-plane array," or FPA) rather than a single element. In this case, we are interested in knowing the degree to which the point in Figure 3.8 maintains its circular shape as the angle of incidence of the collimated beam varies from zero to the cut-off angle, 6° in the current example. Figure 3.16 illustrates the image produced by a collimated beam consisting of 100 000 rays incident to the entrance aperture at an angle of 5°. We see that the point in Figure 3.8 corresponding to on-axis incidence has "spread" to form a feature called "coma." Coma is a common type of distortion associated with reflecting optics. The black dots correspond to rays incident to the focal plane beyond the detector, while the gray dots correspond to rays absorbed by the detector. The blank circle within the coma feature is the shadow of the secondary mirror, while the crescent of dots – some black, some gray – in the lower half of the figure corresponds to stray radiation caused by non-optical rays. An ideal optical prescription would be one for which the optical point-spread function remained circular over a wide range of angles of incidence.

Another important goal of the model-based study is determination of power absorbed by the detector due to secondary infrared emission from the telescope surfaces. This is particularly important if a broadband detector is used in the absence of a band-pass filter. In this case, the detector will be sensitive to temperature fluctuations in, say, the collimator. For the instrument to be of practical use it is important that the designer be able to distinguish power incident to the detector from the telescope structure from power incident to the detector from the scene being observed. Recalling the discussion in Step 1 of Section 3.3, we begin by using the MCRT method to estimate the values of the elements of the radiation distribution factor matrix that represent radiation from the detector to the other surfaces of the instrument. We then invoke the reciprocity property of the distribution factor, Eq. (3.5), to compute the distribution factors from each of the instrument surfaces to the detector. The results of this process are given in Table 3.1.

Table 3.1 Radiation distribution factors from the telescope surfaces to the detector.

Surface, i	$D_{26,i}$	Area, mm^2	ε_i	$D_{i,26}$
1	0.146 23	1809.56	1	0.006 347
2	0.001 65	1809.56	0.9	7.97E-05
3	0.004 42	153.94	0.9	0.002 504
4	0.010 33	17 671.46	0.9	5.1E-05
5	0.000 01	301.59	0.9	1.74E-06
6	0.003 26	1661.90	0.9	0.000 171
7	0.014 67	301.59	0.9	0.004 244
8	0.018 01	17 671.46	0.9	8.89E-05
9	0.000 07	442.96	0.9	1.34E-05
10	0.006 01	1520.53	0.9	0.000 345
11	0.025 24	442.96	0.9	0.004 972
12	0.020 99	17 671.46	0.9	0.000 104
13	0.000 13	578.05	0.9	1.93E-05
14	0.005 36	1385.44	0.9	0.000 338
15	0.013 27	578.05	0.9	0.002 004
16	0.006 01	17 671.46	0.9	2.97E-05
17	0.000 27	706.86	0.9	3.3E-05
18	0.006 77	2981.12	0.9	0.000 198
19	0.000 13	153.94	0.9	7.27E-05
22	0.077 64	122.72	0.9	0.055 212
23	0.545 13	1178.10	0.9	0.040 38
24	0.086 05	7841.42	0.9	0.000 958
25	0.004 41	1178.10	0.9	0.000 327
26	0.003 95	78.54	1	0.003 954
Total	1			

Remember that the radiation distribution factor may be interpreted as a sensitivity, as stated in Eq. (3.1). With this interpretation in mind, perusal of values in the right-hand column of the table leads us to conclude that the influence of temperature fluctuations on Surfaces 2, 4, 5, 8, 9, 13, 16, 17, and 19 will be negligible compared to temperature fluctuations on the remaining surfaces. Specifically, we note that the distribution factors from Surfaces 22 and 23 to the detector exceed those for all other surfaces by more than an order of magnitude. Fortunately, these two surfaces are non-optical surfaces of the primary mirror, which is relatively massive and is buried deep within the instrument. Therefore, it is less likely to undergo temperature excursions during operation. Consistent with the definition of the total, diffuse-specular radiation distribution factor, we can compute the contribution of heat transfer to the detector by any of these surfaces as

$$Q_{i,26} = \varepsilon_i A_i \sigma T_i^{\,4} \mathrm{D}_{i,26}. \tag{3.29}$$

Finally, it is interesting to compute the fraction of power absorbed by the detector for the special case where the entire telescope and the scene being observed are maintained at 300 K. In this case we find that the total detector power is 36.07 mW, while the component of this power attributable to the scene at 300 K is only 0.23 mW. Therefore, a 1% fluctuation in the contribution from the telescope structure would be roughly equivalent to a 100% fluctuation in the signal! This result drives home the need for the instrument temperature to be stabilized during use.

3.6 Use of Radiation Distribution Factors for the Case of Specified Surface Temperatures

Consider the special case where all of the surface elements making up the generic enclosure shown schematically in Figure 3.17 have specified temperatures, and where all of the surface net heat fluxes are unknown. Note that the view of some surface elements may be either partially or fully blocked when viewed from other surface elements. This does not present a problem in the following analysis because partial or full blockage of direct radiation has already been accounted for in the definition and estimation of the distribution factors, as has the effect of surface curvature.

Consistent with the definition of the diffuse-specular graybody radiation distribution factor, we can express the radiation heat flux (W m^{-2})

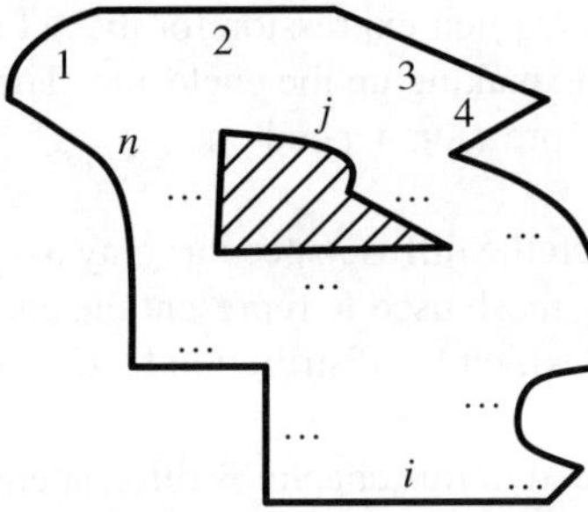

Figure 3.17 Two-dimensional representation of a generic enclosure made up of n surface elements.

absorbed by surface element i due to emission from all of the surface elements making up the enclosure

$$q_{i,a} = \frac{Q_{i,a}}{A_i} = \frac{1}{A_i}\sum_{j=1}^{n} \varepsilon_j A_j \sigma T_j^4 D_{ji}, \quad 1 \le i \le n, \tag{3.30}$$

which, with the introduction of reciprocity, Eq. (3.5), becomes

$$q_{i,a} = \varepsilon_i \sum_{j=1}^{n} \sigma T_j^4 D_{ij}. \quad 1 \le i \le n. \tag{3.31}$$

The radiation heat flux emitted from surface element i is

$$q_{i,e} = \frac{Q_{i,e}}{A_i} = \frac{\varepsilon_i A_i \sigma T_i^4}{A_i} = \varepsilon_i \sigma T_i^4, \quad 1 \le i \le n. \tag{3.32}$$

Then the *net* heat flux *from* surface element i is

$$q_i = q_{i,e} - q_{i,a} = \varepsilon_i \sigma T_i^4 - \varepsilon_i \sum_{j=1}^{n} \sigma T_j^4 D_{ij}, \quad 1 \le i \le n, \tag{3.33}$$

or, with the introduction of the Kronecker delta

$$\delta_{ij} \equiv \begin{cases} 1, & i = j \\ 0, & i \ne j \end{cases}, \tag{3.34}$$

Eq. (3.33) can be written

$$q_i = \varepsilon_i \sum_{j=1}^{n} \sigma T_j^4 (\delta_{ij} - D_{ij}) \quad 1 \le i \le n. \tag{3.35}$$

Equation (3.35) is an explicit expression for the net heat flux from all of the n surface elements making up the enclosure. The accuracy of results obtained using this expression depend on

1. the degree to which the diffuse-specular gray assumption is valid,
2. the fineness of the mesh used to represent the enclosure, and
3. the accuracy with which the distribution factors have been estimated.

The approach presented in this chapter is often adequate for preliminary design studies. Chapter 4 deals with improvements in accuracy that can be achieved by including surface models that account for directionality and/or wavelength dependence, and a formalism allowing statement of the uncertainty and corresponding confidence interval is developed in Chapter 7.

3.7 Use of Radiation Distribution Factors When Some Surface Net Heat Fluxes Are Specified

We now address the situation where surface elements 1, 2, … , N have specified net heat fluxes and surfaces $N+1$, $N+2$, … , n have specified temperatures. To begin, let us consider the special case where $N=1$; that is, where only the first of n surface elements has a specified net heat flux, with the remaining surfaces having specified temperatures. In this case Eq. (3.35) may be written

$$q_1 = \varepsilon_1 \left[(1 - D_{11})\sigma T_1^4 - \sum_{j=2}^{n} \sigma T_j^4 D_{ij} \right]. \tag{3.36}$$

Equation (3.36) can be solved explicitly for the unknown surface temperature T_1 in terms of the known surface net heat flux q_1 and the known surface temperatures; that is,

$$\sigma T_1^4 = \frac{1}{1 - D_{11}} \left(\frac{q_1}{\varepsilon_1} + \sum_{j=2}^{n} \sigma T_j^4 D_{ij} \right). \tag{3.37}$$

In the more general case where several surface elements have specified net heat fluxes we can rearrange Eq. (3.35) to obtain

$$q_i + \varepsilon_i \sum_{j=N+1}^{n} \sigma T_j^4 D_{ij} = \varepsilon_i \sum_{j=1}^{N} \sigma T_j^4 (\delta_{ij} - D_{ij}), \quad 1 \le i \le N. \tag{3.38}$$

Equation (3.38) represents N equations in the N unknown surface temperatures in terms of the N known surface net heat fluxes and the $n-N$ known surface temperatures. It can be rewritten symbolically as

$$\Theta_i = \Psi_{ij}\Omega_j, \tag{3.39}$$

where

$$\Theta_i = q_i + \varepsilon_i \sum_{j=N+1}^{n} \sigma T_j^4 D_{ij}, \quad 1 \leq i \leq N, \tag{3.40}$$

is a known vector,

$$\Psi_{ij} = \varepsilon_i(\delta_{ij} - D_{ij}), \quad 1 \leq i \leq N, \quad 1 \leq j \leq N, \tag{3.41}$$

and

$$\Omega_j = \sigma T_j^4, \quad 1 \leq j \leq N, \tag{3.42}$$

is an unknown vector whose elements are sought. We obtain the unknown surface temperatures by inverting the matrix defined by Eq. (3.41) and then using it to operate on the vector defined by Eq. (3.40); that is,

$$\Omega_i = [\Psi_{ij}]^{-1}\Theta_j, \quad 1 \leq i \leq N. \tag{3.43}$$

The unknown surface net heat fluxes are then computed using Eq. (3.35) applied over the range $N+1 \leq i \leq n$.

Problems

3.1 Derive Eq. (3.4).

3.2 Derive Eq. (3.5).

3.3 Derive Eq. (3.6).

3.4 Modify the logic block diagram in Figure 3.1 for the case where the power contained by the energy bundle decreases by an amount proportional to the absorption coefficient with each reflection.

3.5 Use the MCRT method with 10 000 rays to demonstrate that Eq. (3.23) produces a uniform distribution of intersections with the unit sphere in Figure 3.4. Explain why this is the desired result (see Problem 1.13).

3.6 Derive Eqs. (3.25) and (3.26).

3.7 Derive Eq. (3.27).

3.8 Compose the Matlab code required to implement the logic block diagram in Figure 3.1 for modeling radiant exchange in the instrument shown in Figure 3.7. The surface numbers in Figure 3.6 are indicated by the diagonal lines. Use appropriate descriptive variable names. Organize the code into clearly labeled blocks that perform each of the steps.

3.9 Consider the wedge-shaped cavity depicted in Figure 1.15. Suppose the aperture is W μm wide and is centered at $y_{max}/3$. The remaining dimensions and the angle α are as given in Problem 1.19. If the walls are gray with the absorption coefficient α (not to be confused with the angle α) and the specularity ratio $\rho^s/(1 - \alpha)$ both equal to 0.9, what is the distribution factor of the aperture to itself, D_{aa}? In other words, what fraction of the entering power escapes from the cavity? Assume that the power entering the cavity through the aperture is diffusely distributed and that the energy bundles retain all of their power until they are either absorbed in the cavity walls or escape through the aperture. Let the aperture width W take on values of 10, 20, and 40 μm. For each aperture width compute the value of D_{aa} as a function of the number N_a of energy bundles entering through the aperture. Let the number of rays traced be 10^3, 10^4, 10^5, and 10^6. Plot your results on log–log coordinates. [Hint: Remember to take advantage of the reciprocity relation, Eq. (3.5), where appropriate.]

3.10 Suppose the temperature of the interior walls of the wedge-shaped cavity in Problem 3.9 is 300 K. Use the Stefan–Boltzmann relation $Q = \varepsilon\sigma T^4$, where $\varepsilon = \alpha$, to calculate the flux (W m^{-2}) emitted from the surfaces.

(a) Trace a total of one million rays and plot the amount of power in milliwatts escaping through the aperture as a function of its width over the range $0 \leq W \leq y_{max}$.

(b) Plot scatter diagrams of the spatial distribution of rays incident to a screen 20 mm in front of the slit. Let the screen area be 100 × 100 mm.

3.11 Figure 3.18 shows a sphere centered halfway between two parallel rectangular plates. The upper plate (Surface 3) has the same dimensions as, and is centered on, the lower plate (Surface 2). Surface 2 is diffuse with an emissivity of ε_2, and Surface 3 is specular with a reflectivity of ρ_3^s. The sphere (Surface 1) is diffuse with an emissivity of ε_1. The surroundings are black. Compose a Matlab

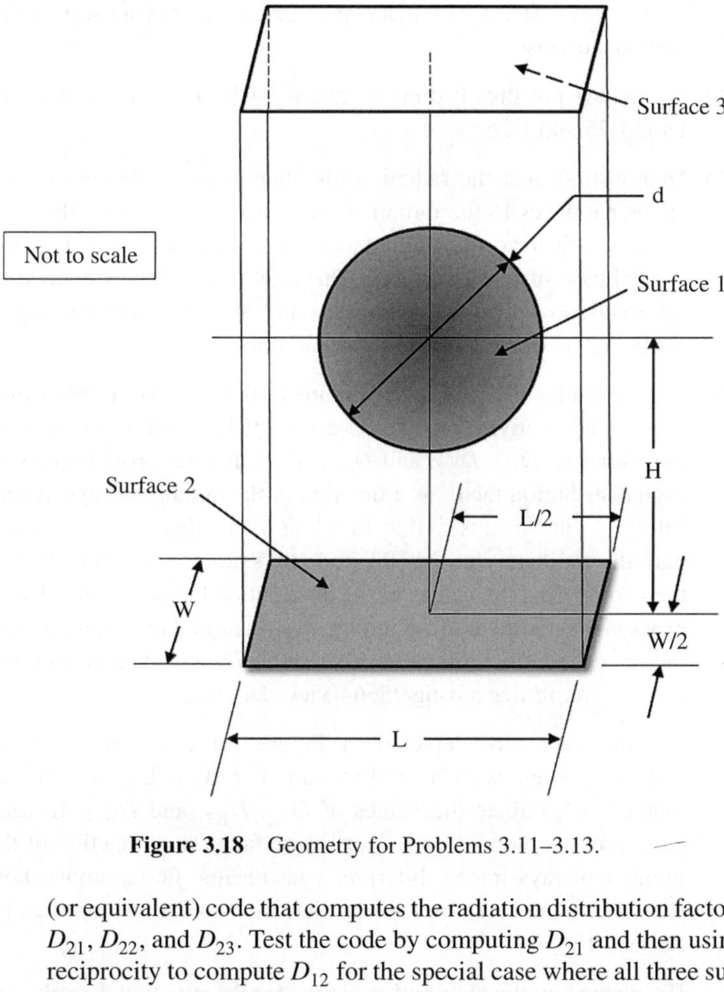

Figure 3.18 Geometry for Problems 3.11–3.13.

(or equivalent) code that computes the radiation distribution factors D_{21}, D_{22}, and D_{23}. Test the code by computing D_{21} and then using reciprocity to compute D_{12} for the special case where all three surfaces are black and $L = W = 2d = 2H$. Plot the value of D_{12} as a function of the number of rays traced. Does the solution converge to the expected value? What is the expected value? Plot the distribution of absorbed energy bundles on the surfaces.

3.12 Use the code developed in Problem 3.11 to study the behavior of D_{22} for a range of values of ε_1, ε_2, and ρ_3 with $L = W = 2d = 2H$. Interpret your results. Do they make sense?

3.13 Use the code developed in Problem 3.11 to study the behavior of D_{22} for a range of values of $2H/d > 1$, with $L = W = 2H$. For this study let $\varepsilon_1 = 0.5$, $\varepsilon_2 = 0.8$, and $\rho_3 = 0.9$. Interpret your results. Do they make sense?

3.14 If you have not already done so, return to Chapter 1 and work Problems 1.25 and 1.26.

3.15 Demonstrate that the radiation distribution factor defined in this chapter reduces to the radiation view factor (also called the configuration factor or the angle factor by various authorities) for the special case of black surfaces. The view factor $F_{1\text{–}2}$ is defined as the fraction of diffuse radiation leaving Surface 1 and arriving at Surface 2 by direct (line-of-sight) radiation.

3.16 For the enclosure depicted in Figure 1.16, let the walls be diffuse with an absorptivity of 0.55. Use the MCRT method to estimate the values of $D_{1,2}$, $D_{1,7}$, and $D_{10,4}$. In each case, provide plots of each distribution factor as a function of the number of rays traced. Interpret your results. [Hint: Problem 3.15 suggests that you can partially validate your MCRT code by setting $\alpha = 1$ (black) and then comparing the values of the distribution factors obtained with exact values of the corresponding view factors. For a Matlab view factor calculation function visit https://www.mathworks.com/matlabcentral/fileexchange/5664-view-factors.]

3.17 For the enclosure depicted in Figure 1.16, let the walls be diffuse-specular with $\alpha = 0.55$ and $\rho^s = 0.8$. Use the MCRT method to estimate the values of $D_{1,2}$, $D_{1,7}$, and $D_{10,4}$. In each case, provide plots of each distribution factor as a function of the number of rays traced. Interpret your results. For example, how sensitive is the heat absorbed by Surface 4 to the heat emitted by Surface 10?

3.18 Estimate all of the distribution factors for the situation described in Problem 3.16. Let Surfaces 1–4 be maintained at a temperature of 800 K and Surfaces 5–8 be maintained at a temperature of 200 K, with the remaining surfaces insulated. Compute:
(a) the temperatures (K) of the insulated surfaces and
(b) the net heat transfer (W) from Surfaces 1–8.

3.19 If you have not already done so, work Problems 2.26 and 2.28. Then, letting Surfaces 1–4 be pure platinum and the remaining surfaces be black glass, compute the corresponding distribution factor matrix using the surface optical models obtained from working Problems 2.26 and 2.28.

3.20 Assume that the surface optical models obtained in Problem 3.19 are independent of temperature and wavelength (graybody assumption). Let Surfaces 1–4 be maintained at a temperature of 800 K and Surfaces 5–8 be maintained at a temperature of 200 K, with the remaining surfaces insulated. Compute:

(a) the temperatures (K) of the insulated surfaces and
(b) the net heat transfer (W) from Surfaces 1–8.

Reference

1. Kowsary, F. (1999). A computationally efficient method for Monte Carlo simulation of diffuse radiant emission or reflection (Technical Note). *International Journal of Heat and Mass Transfer* 42: 193–195.

4

Extension of the MCRT Method to Non-Diffuse, Non-Gray Enclosures

The assumption of diffuse emission and diffuse-specular reflection in gray enclosures is adequate for many applications encountered in engineering practice. This is especially true in the conceptual and preliminary design stages. However, final design and performance studies often require the use of bidirectional surface optical models restricted to a specified wavelength interval. In this chapter we develop the concept of the band-averaged spectral radiation distribution factor, which is the most general tool available for treating radiant exchange among surfaces, but first we learn through extended examples how to compose and use models to simulate the optical behavior of non-diffuse surfaces.

4.1 Bidirectional Spectral Surfaces

Experience confirms that reflection from a surface is generally neither diffuse nor specular. Rather, at a given wavelength the distribution of reflected energy depends on the mechanical and chemical preparation

The Monte Carlo Ray-Trace Method in Radiation Heat Transfer and Applied Optics, First Edition. J. Robert Mahan.

This Work is a co-publication between John Wiley & Sons Ltd and ASME Press.
Companion website: www.wiley.com/go/mahan/monte-carlo

of the surface and on the direction of incidence. Diffuse and specular reflections represent the two extremes of bidirectional reflectivity. Both extremes may be approached but rarely achieved in practice. Various theories have been proposed for predicting bidirectional spectral reflection and directional spectral emission and absorption for generic surfaces. The interested reader is referred to Chapter 4 in Ref. [1], where this topic is pursued in more detail. However, metrology remains the only sure path to surface models that accurately capture the directional spectral behavior of real surfaces.

A simple two-component model for directional reflectivity was introduced in Chapter 1 (Figure 1.9), where it is suggested that a directional reflection pattern can be somewhat approximated as a suitably weighted combination of spectral and diffuse reflection. Different versions of this approximation would have to be applied for each wavelength of interest, using different weight factors for each wavelength. Any success this approach might have would be due in large measure to the fact that the distribution of radiant energy within an enclosure is governed by integral equations rather than by differential equations. Integration at least partially "averages out" positive and negative excursions from reality, as illustrated in Figure 4.1. Inspection of the figure reveals that, even though

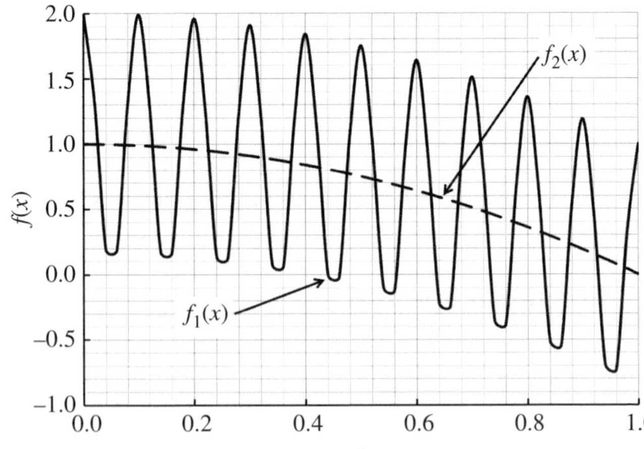

Figure 4.1 The areas under the two curves are about the same even though the curves themselves are different.

$f_1(x)$ is a more detailed and presumably more accurate description of local behavior than $f_2(x)$, it may nonetheless be true, to an acceptable approximation, that

$$\int_0^1 f_1(x)dx \cong \int_0^1 f_2(x)dx. \tag{4.1}$$

That is, the area under the two curves may be essentially the same.

Departures from reality of the simple model presented in Chapter 1 often create the need for a more accurate approach. In the Matlab environment it would be convenient to have available a *function* of the form

$$\textit{function}[\textit{reflectivity}] = \text{REFLECTION}\,(\textit{theta\_incident}, \textit{theta\_reflected}, \textit{phi\_reflected}),$$

where *reflectivity* is the bidirectional reflectivity and the angles *theta* (ϑ) and *phi* (φ) are illustrated in Figure 4.2. Note that the input variable list does not include the variable name *phi_incident*. This is because most surfaces of practical engineering interest are **isotropic**; that is, they produce a reflection pattern that is invariant with the azimuth angle of incidence. Before invoking the function, random values of *theta_reflected* and *phi_reflected* would first be obtained corresponding to a diffuse reflection, using Eq. (1.32). The function would then return the fraction of the incident power, *reflectivity*, carried by a ray reflected in that direction. Creation and use of a suitable bidirectional spectral reflection function is exemplified in the following case studies.

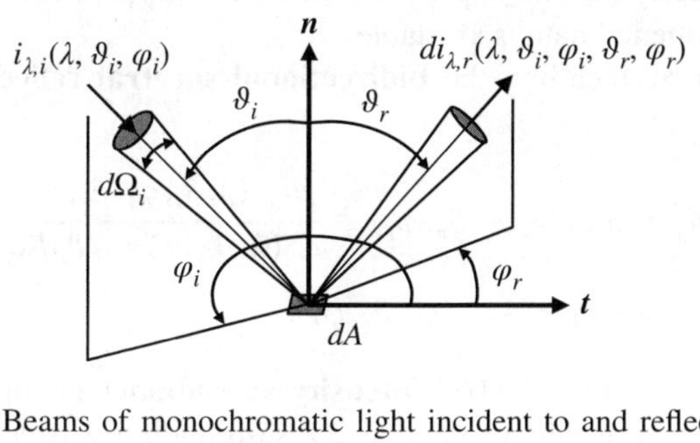

Figure 4.2 Beams of monochromatic light incident to and reflected from an area element dA.

4.2 Principles Underlying a Practical Bidirectional Reflection Model

Useful bidirectional reflection models are generally based on measurements, although they are frequently informed by theory. As we learned in Sections 2.12 and 2.13, the optical behaviors of electrically non-conducting (dielectric) and electrically conducting (metal) surfaces are fundamentally different. In general, metal surfaces are strong specular reflectors while dielectric surfaces tend to be weak diffuse reflectors. In both cases the directional distribution of reflected radiation is known to be strongly influenced by the topography, chemical state, and degree of contamination of the surface. With the exception of certain optical components (such as mirrors, lenses, and filters), it is unlikely that a bidirectional spectral reflectivity model based entirely on theory would accurately represent the optical behavior of a surface of practical engineering interest. Therefore, in cases where high accuracy is required, a successful surface optical model must be at least semiempirical if not based entirely on measurements of the optical behavior of the surface to be modeled.

In this chapter we first demonstrate the application of semiempirical approaches for two surface coatings engineered to exhibit specific – and somewhat unique – optical behaviors. In the first example we consider a highly absorptive commercial coating whose small component of reflectivity is highly directional to the point of being almost specular, and in the second example we consider another commercial coating that is highly reflective but whose reflectivity is nearly diffuse. Both of these coatings are widely used in optical applications requiring an unusual combination of both metallic and dielectric behaviors. We then follow up by presenting a completely general approach suitable for applications where a full set of experimental data is available.

We begin by recalling the **bidirectional spectral reflectivity** from Chapter 2,

$$\begin{aligned}\rho''_\lambda = \rho(\lambda, \vartheta_i, \varphi_i, \vartheta_r, \varphi_r) &\equiv \frac{di_{\lambda,r}(\lambda, \vartheta_i, \varphi_i, \vartheta_r, \varphi_r)}{i_{\lambda,i}(\lambda, \vartheta_i, \varphi_i)\cos\vartheta_i d\Omega_i} \\ &\equiv BRDF(\lambda, \vartheta_i, \varphi_i, \vartheta_r, \varphi_r). \qquad (4.2)\end{aligned}$$

In Eq. (4.2), i_λ is the **spectral intensity**, or **radiance** ($\text{W m}^{-2} \cdot \text{sr} \cdot \mu\text{m}$), of radiation in wavelength interval $\Delta\lambda$ surrounding wavelength λ, and the subscripts i and r are associated with the incident and reflected beams, respectively. The incident and reflected zenith angles, ϑ_i and ϑ_r, the incident and reflected azimuth angles, φ_i and φ_r, and the incident differential solid angle $d\Omega_i$ are illustrated in Figure 4.2. In contemporary technical literature the bidirectional reflectivity is often referred to as the **bidirectional reflectance distribution function**, or *BRDF*. Nearly all surfaces of practical engineering interest are **isotropic**, which means that the *BRDF* is independent of the azimuth angle of incidence, φ_i. In the following analysis we suppress φ_i from the various argument lists, thereby limiting the results obtained to isotropic surfaces.

A related quantity is the **monochromatic bidirectional reflectance factor**, or *BRF*, which compares the flux in wavelength interval $d\lambda$ about λ reflected from a real surface in direction (ϑ_r, φ_r) due to a flux incident at angle ϑ_i with the mean flux averaged over the reflected 2π-space,

$$BRF(\lambda, \vartheta_i, \vartheta_r, \varphi_r) \equiv \frac{d\Phi_r(\lambda, \vartheta_i, \vartheta_r, \varphi_r)}{\frac{1}{2\pi}\int_{2\pi_r} d\Phi_r(\lambda, \vartheta_i, \vartheta_r, \varphi_r) d\Omega_r} = \frac{d\Phi_r(\lambda, \vartheta_i, \vartheta_r, \varphi_r)}{\langle d\Phi_r(\lambda, \vartheta_i)\rangle}. \tag{4.3}$$

Because $d\Phi_r = di_r \cos(\theta_r)$ and $\langle d\Phi_r\rangle = \langle di_r\rangle \cos(\theta_r)$, Eq. (4.3) can be rewritten

$$BRF(\lambda, \vartheta_i, \vartheta_r, \varphi_r) = \frac{di_r(\lambda, \vartheta_i, \vartheta_r, \varphi_r)}{\langle di_r(\lambda, \vartheta_i)\rangle} = \frac{di_r(\lambda, \vartheta_i, \vartheta_r, \varphi_r)}{di_{r,diffuse}(\lambda, \vartheta_i)}, \tag{4.4}$$

where $di_{r,\ diffuse}(\lambda, \vartheta_i)$ is the equivalent reflected intensity from a diffuse surface. Thus we see that $BRF = 1.0$ for the special case of a diffusely reflecting surface.

When measuring the *BRF* it is usual to illuminate the surface with a diffuse source, that is with a source whose intensity $i_{i,\ diffuse}(\lambda)$ is independent of angle of incidence. With this restriction it is convenient to rewrite Eq. (4.4) as

$$\begin{aligned} BRF(\lambda, \vartheta_i, \vartheta_r, \varphi_r) &= \frac{di_r(\lambda, \vartheta_i, \vartheta_r, \varphi_r)}{di_{i,diffuse}(\lambda)} \Bigg/ \frac{di_{r,diffuse}(\lambda)}{di_{i,diffuse}(\lambda)} \\ &= \frac{1}{\rho(\lambda)} \frac{di_r(\lambda, \vartheta_i, \vartheta_r, \varphi_r)}{di_{i,diffuse}(\lambda)}, \end{aligned} \tag{4.5}$$

where $\rho(\lambda) \equiv i_{r,diffuse}(\lambda)/i_{i,diffuse}(\lambda)$ is the (bi)hemispherical spectral reflectivity of the surface for the special case of diffuse radiation. The factor $1/\rho(\lambda)$ in Eq. (4.5) accounts for the "lossiness" of the surface, while the factor $i_r(\lambda, \vartheta_i, \vartheta_r, \varphi_r)/i_{i,ideal}(\lambda)$ accounts for its directionality. It is left as an exercise to show that the *BRF* is related to the *BRDF* by

$$BRF = \pi BRDF. \tag{4.6}$$

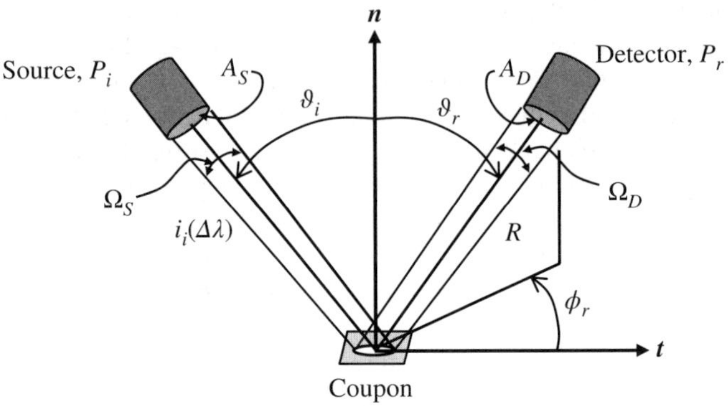

Figure 4.3 Schematic representation of an apparatus for measuring the *BRDF* of a coupon.

The *BRDF* and the *BRF* are typically measured using the apparatus illustrated schematically in Figure 4.3. The azimuth angle of reflection φ_r is measured anti-clockwise with respect to the **plane of incidence**, defined as the plane containing the incident ray, the surface normal $\boldsymbol{n}$, and the surface tangent $\boldsymbol{t}$. A detector whose effective aperture area is A_D is oriented normal to a line of length R extending from the center of the coupon under test to the detector. The reflected power P_r sensed by the detector is then $i_r(\Delta\lambda, \vartheta_r, \varphi_r)A_D\Omega_D$, where $\Omega_D = A_D/R^2$ is the solid angle subtended by A_D at the center of the coupon. The symbol $\Delta\lambda$ refers to the effective wavelength interval of the light from the source. Then measurements of P_r are obtained at a manifold of detector locations (ϑ_r, φ_r) corresponding to illumination by a source of power $P_S = i_S A_S \Omega_S$ incident at each angle ϑ_i. In principle the lossiness of the surface, represented by $1/\rho(\Delta\lambda)$, may be determined by computing

$$\rho(\Delta\lambda) = \frac{\int_{\frown_r} \Phi_r(\Delta\lambda, \vartheta_r, \varphi_r)d\Omega_r}{\int_{\frown_i} \Phi_i(\Delta\lambda, \vartheta_i, \varphi_i)d\Omega_i} \approx \frac{\sum_n \frac{P_{r,n}(\Delta\lambda,\vartheta_r,\varphi_r)}{(2\pi R^2)}}{\pi i(\Delta\lambda)}, \tag{4.7}$$

where $\Phi_r(\Delta\lambda, \vartheta_r, \varphi_r)$ is the reflected flux ($\mathrm{W\,m^{-2}}$) in direction (ϑ_r, φ_r) and $\Phi_i(\Delta\lambda, \vartheta_i, \varphi_i)$ is the incident flux from direction (ϑ_i, φ_i). The summing index n in Eq. (4.7) refers to the number of detector positions required to effectively "cover" the hemispherical surface in whch the detector is displaced. The accuracy of the estimate increases with the number of positions $n = 2\pi R^2/A_D$. In practice, $\rho(\Delta\lambda)$ is usually measured separately using an integrating sphere. The role of the integrating sphere in optical metrology is investigated further in Section 4.5.

In terms of the parameters defined above and in Figure 4.3, Eq. (4.5) can be written

$$BRF(\Delta\lambda, \vartheta_i, \varphi_i, \vartheta_r, \varphi_r) = \frac{1}{\rho(\Delta\lambda)} \frac{P_r(\Delta\lambda, \vartheta_i, \varphi_i, \vartheta_r, \varphi_r)/(A_D\Omega_D \cos\vartheta_r)}{P_i(\Delta\lambda\vartheta_i, \varphi_i,)/(A_S\Omega_S \cos\vartheta_i)}. \tag{4.8}$$

When using Eq. (4.8), it is convenient to configure the apparatus so that $A_D\ \Omega_D = A_S\ \Omega_S$, yielding

$$BRF(\Delta\lambda, \vartheta_i, \vartheta_r, \varphi_r) = \frac{1}{\rho(\Delta\lambda)} \frac{P_r(\Delta\lambda, \vartheta_r, \varphi_r)/\cos\vartheta_r}{P_i(\Delta\lambda)/\cos\vartheta_i}. \tag{4.9}$$

We classify the two general approaches for treating bidirectional reflectivity in the MCRT environment as direct and indirect. In the direct approach, unprocessed bidirectional reflectivity data are stored in and accessed on demand from a large data file. This approach, while inherently more accurate than the indirect approach, places significantly higher demands on both experimental and computing resources. In the indirect approach, sparse experimental data are preprocessed into a relatively compact bidirectional reflectivity model, which is then inserted into the logical flow of the MCRT code. This approach usually involves various approximations aimed at economizing on computer speed and storage. We begin by considering two examples of this latter approach.

4.3 First Example: A Highly Absorptive Surface Whose Reflectivity is Strongly Specular

Aeroglaze® Z302 [2] is a polyurethane-based paint whose absorptivity typically exceeds 90% in the visible part of the spectrum, depending on the coating thickness. It is unique in that the reflected component of radiation is mostly specular. Its special properties make it the coating of choice for many aerospace and optical applications where a surface must be an exceptionally efficient absorber but where diffuse reflection is undesirable. A typical application is the interior surface of a blackbody cavity used as a calibration target. In this case the cavity geometry would be such that several specular reflections would occur before an incident ray could escape, and any diffuse component of reflectivity present would diminish the effectiveness of the design because it would allow some power to escape the cavity with each reflection. Such diffuse "leaks" can be significant when the effective emissivity of the cavity must be unity to better than three nines.

Prokhorov and Prokhorova [3] describe a three-component semiempirical model based on their own measurements of the *BRDF* of Z302 at a wavelength of $\lambda = 10.6$ μm. We have used the same data, represented by the symbols in Figure 4.4, to derive a purely empirical four-component model, represented by the curves in the figure. Both the Prokhorov and Prokhorova model and our model are in excellent agreement with the measurements. Our four-component model [4] has the form

$$BRDF = \rho_1'' + \rho_2'' + \rho_3'' + \rho_4'', \tag{4.10}$$

where

$$\rho_n'' = A_n \frac{1}{\sqrt{2\pi}\sigma_n} e^{-(\vartheta_v - \vartheta_i)^2/2\sigma_n^2} + O_n, \quad n = 1, 2, 3, 4. \tag{4.11}$$

In Eq. (4.11), ϑ_i and ϑ_v are the incidence and viewing angles shown in the inset in Figure 4.4, and A_n, σ_n, and O_n are empirical curve-fitting parameters. The form of Eq. (4.11) is recognizable as the normal probability distribution function multiplied by a scaling factor A_n and shifted

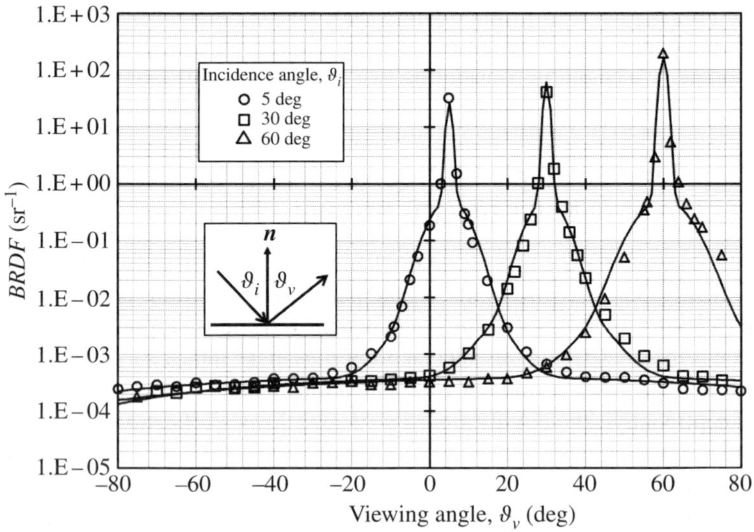

Figure 4.4 Empirical model [4] (curves) to Prokhorov and Prokhorova data [3] (symbols).

in amplitude by an offset O_n. In practice the additive offsets for the four values of n are gathered into a single constant. The fit illustrated in Figure 4.4 was obtained by defining the standard deviation

$$\sigma_n = \frac{1}{\sqrt{2\pi b_n \vartheta_i}} \tag{4.12}$$

and the multiplicative constant

$$A_n = \frac{B_n}{b_n \vartheta_i}, \tag{4.13}$$

where the coefficients B_n and b_n and the offset are given in Table 4.1. We see that the coefficients B_1, B_2, B_3, and the offset are constants, while the coefficients b_1, b_2, b_3, b_4, and B_4 are functions of the incidence angle, ϑ_i. The variations of these latter parameters with incidence angle are illustrated in Figures 4.5 and 4.6, which include the corresponding fitting equations.

The similarity of form among the fits for b_n suggests that they are governed by a common physical principle. Note that the values for b_n are unbounded at $\vartheta_i = 0$. Therefore, since values of ϑ_i approaching zero may occur in a practical ray trace, it is necessary to artificially limit these parameters below a certain threshold value of ϑ_i. In the case at hand we use the rule that if $\vartheta_i < 1°$, $b_n = b_n(1°)$. The overall error introduced by this rule is negligible because of the relatively small number of incidences of $\vartheta_i < 1°$, but especially because the value of the exponential in Eq. (4.11) rapidly approaches 1.0 as ϑ_i decreases below 1°. The fit for the variation of the coefficient B_4 with incidence angle is given in Figure 4.6. The choice of a log-linear fitting function is justified because only three observations are available. A more elegant fitting function might produce a better fit, but with only three data points it would place too much weight on a single observation.

We now have an analytical tool for characterizing the bidirectional reflectivity, or *BRDF*, of Z302 near 10.6 μm. However, this model

Table 4.1 Fitting parameter values for the model shown in Figure 4.4 [4].

ϑ_i	b_1	B_1	Offset, O	b_2	B_2	b_3	B_3	b_4	B_4
5	0.000250	0.005	0.0046	0.0100	0.01	0.0200	0.5	0.12	32.1
30	0.000040	0.005	0.0046	0.0015	0.01	0.0035	0.5	0.02	40.8
60	0.000015	0.005	0.0046	0.0016	0.01	0.0015	0.5	0.01	195

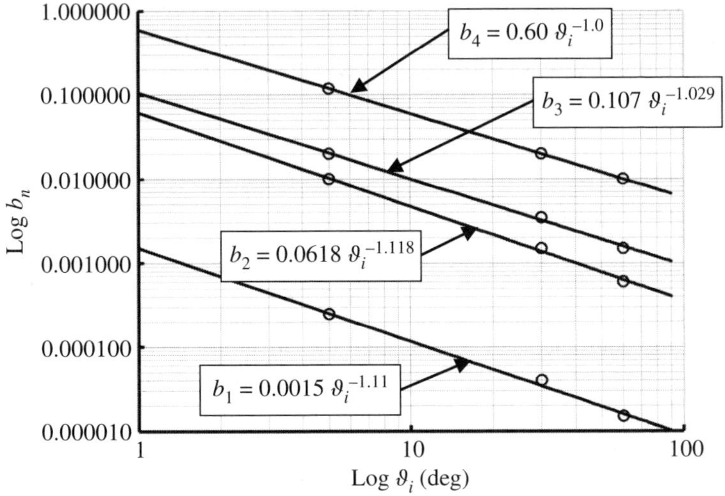

Figure 4.5 Models for b_n coefficients (symbols are values from Table 4.1) [4].

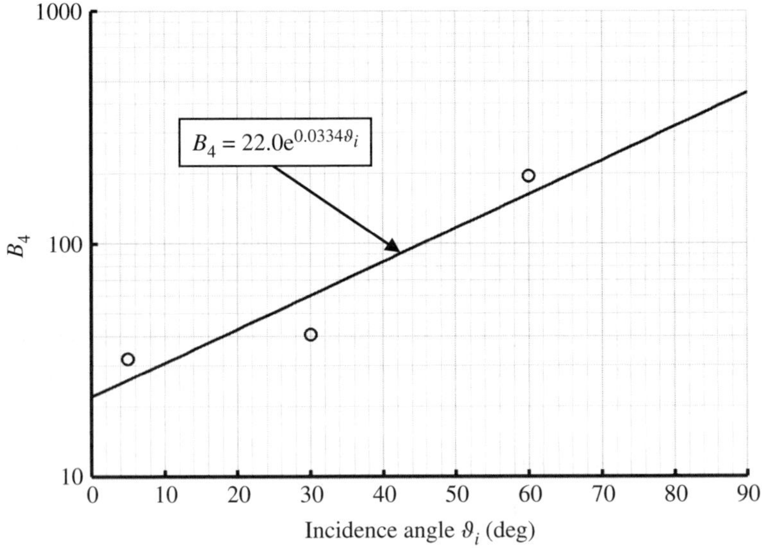

Figure 4.6 Model for B_4 coefficient (symbols are values from Table 4.1) [4].

describes the surface optical behavior only for viewing angles in the plane of incidence. How do we predict the reflectivity at viewing angles out of the plane of incidence? Following Prokhorov and Prokhorova, we assume axisymmetry in the directional reflection pattern about the axis running from the point of reflection in the direction of the viewing zenith angle, ϑ_v. Subject to this reasonable assumption, the model represented by Eqs. (4.10) and (4.11) is valid for all viewing angles if $\vartheta_v - \vartheta_i$ in Eq. (4.11) is replaced by γ, where

$$\gamma = (180.0/\pi)\cos^{-1}[\boldsymbol{v}_r \cdot \mathbf{v}_v]\ (\text{deg}). \tag{4.14}$$

In Eq. (4.14)

$$\boldsymbol{v}_r = (x_r - x_1)\,\boldsymbol{i} + (y_r - y_1)\boldsymbol{j} + (z_r - z_1)\,\boldsymbol{k} \tag{4.15}$$

and

$$\boldsymbol{v}_v = (x_v - x_1)\,\boldsymbol{i} + (y_v - y_1)\boldsymbol{j} + (z_v - z_1)\,\boldsymbol{k}, \tag{4.16}$$

where (x_1, y_1, z_1) is the point on the surface where the ray is incident, (x_r, y_r, z_r) is a point on the surface of a unit hemisphere intersected by a diffusely reflected ray, and (x_v, y_v, z_v) is the point of the unit hemisphere intersected by a ray at the viewing angle $\vartheta_v = \vartheta_i$ in Figure 4.4. The geometry corresponding to Eqs. (4.14)–(4.16) is illustrated in Figure 4.7.

The various coefficients comprising this model are of course particular to Z302 at 10.6 μm applied to the thickness of the coating actually tested by Prokhorov and Prokhorova. In general, the values of these coefficients can be expected to vary with both wavelength and coating

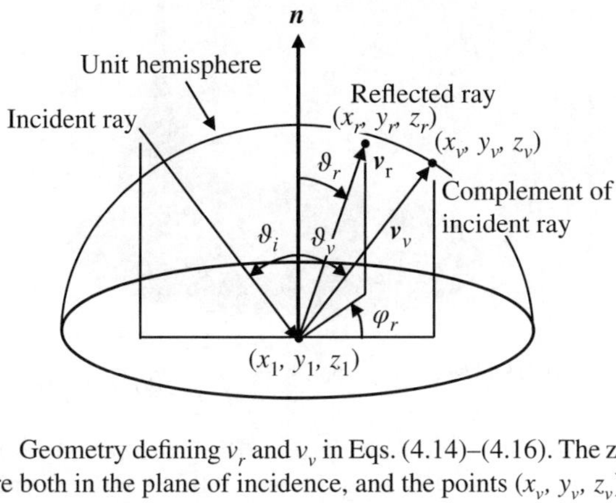

Figure 4.7 Geometry defining v_r and v_v in Eqs. (4.14)–(4.16). The zenith angles ϑ_i and ϑ_v are both in the plane of incidence, and the points (x_v, y_v, z_v) and (x_r, y_r, z_r) both lie in the surface of the unit hemisphere [4].

thickness, although the variation with thickness will be minor beyond a minimum thickness. Prokhorov and Prokhorova [3] report that the wavelength dependence of Z302 is rather weak in the interval surrounding 10.6 μm, and they suggest that modeling parameters determined at this wavelength can probably be used across the 8–12-μm range with acceptable error. It is difficult to overstate the need for conducting a serious experimental campaign aimed at characterizing the optical behavior of the actual surface coatings to be used in an application requiring a high level of accuracy.

Figure 4.8 represents the *BRDF* distribution corresponding to 100 000 rays created by the splitting of a single ray incident at $(x, y, z) = (0, 0, 0)$ at an angle of 45° according to the four-component bidirectional reflectivity model. The illuminated sample lies in the x,y-plane. All rays carry the same power.

Once the bidirectional reflectivity, or *BRDF*, has been computed, the **directional absorptivity** may be determined as

$$\alpha(\vartheta_i) = 1 - \rho(\vartheta_i), \tag{4.17}$$

where

$$\rho(\vartheta_i) = DHR = \int_0^{2\pi} \int_0^{\pi/2} \rho(\vartheta_i, \vartheta_r, \varphi_r) \cos \vartheta_r \sin \vartheta_r d\vartheta_r d\varphi_r \tag{4.18}$$

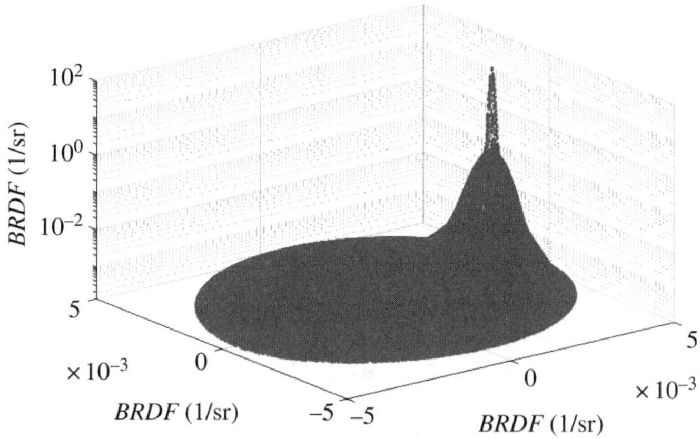

Figure 4.8 Scatter diagram showing the *BRDF* distribution resulting when a single ray incident at 45° is split into 100 000 rays according to the four-component model [4].

is the directional-hemispherical reflectivity (*DHR*). Prokhorov and Prokhorova report a value of about 0.051 (read from figure 4 in Ref. [3]) for the directional-hemispherical reflectivity of their sample for an incidence angle of 8°. This value is plotted in Figure 4.9, which shows the values of directional-hemispherical reflectivity obtained using the four-component model for $\rho(\vartheta_i, \vartheta_r, \varphi_r)$ when the integration implied by Eq. (4.18) is carried out using Eq. (4.19). The agreement between the directly measured value and the value obtained using the model – within 3% – is remarkably good considering that the two methods for determining $\rho(\vartheta_i)$ are based on two different independent experiments, with data from one of them interpreted using a complex model. The horizontal dashed line in Figure 4.9 is drawn at the level of the (bi)hemispherical reflectivity $\rho(\lambda)$ introduced in Eq. (4.5) as estimated using Eq. (4.7). The variation with angle of incidence of the directional absorptivity of Z302 at a wavelength of 10.6 μm, based on the four-component model, is shown in Figure 4.10.

We now consider an incident ray carrying power P_i that is partially reflected from a directionally reflecting surface whose bidirectional reflectivity is $\rho(\vartheta_i, \vartheta_r, \varphi_r)$. The directional-hemispherical reflectivity

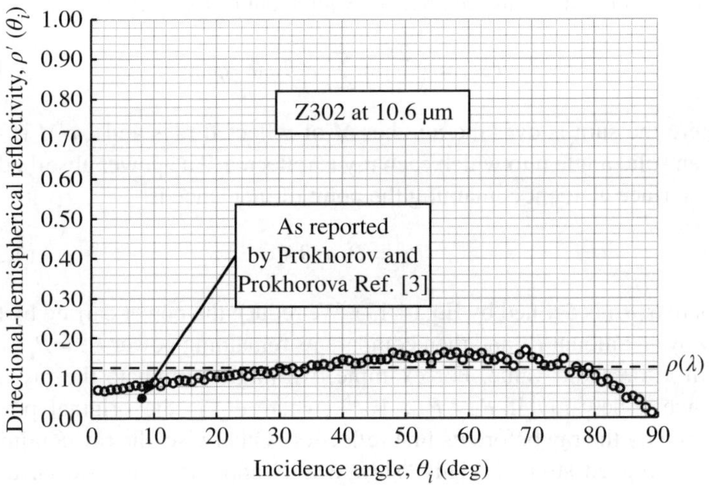

Figure 4.9 The variation of the directional-hemispherical reflectivity of Z302 at $\lambda = 10.6\,\mu m$ computed using Eq. (4.18) with the four-component model for $\rho(\vartheta_i, \vartheta_r, \varphi_r)$ [4].

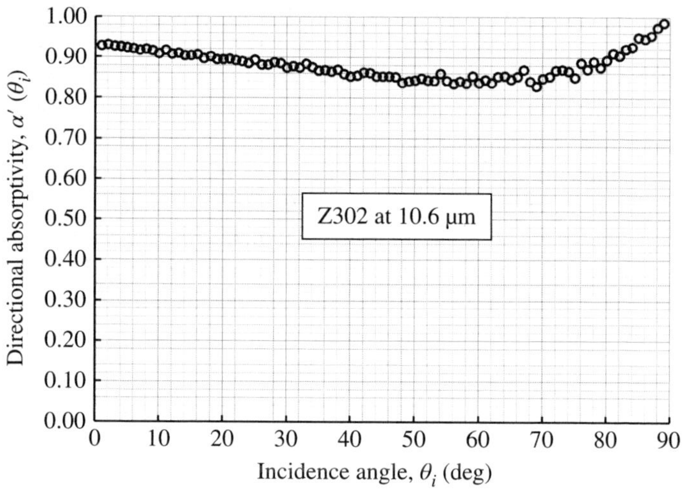

Figure 4.10 The variation of the directional absorptivity of Z302 at $\lambda = 10.6\ \mu\text{m}$ computed using Eqs. (4.17) and (4.18) with the four-component model for $\rho(\vartheta_i, \vartheta_r, \varphi_r)$ [4].

$\rho(\vartheta_i)$ is computed using the MCRT equivalent of Eq. (4.18),

$$\rho(\vartheta_i) = \left(\frac{2\pi}{M}\right) \sum \rho(\vartheta_i, \vartheta_r, \varphi_r), \tag{4.19}$$

where the sum is over the number M of reflected rays and $2\pi/M$ is the mean solid angle into which each ray is reflected. The power absorbed in the surface element containing the point of incidence is

$$P_a = P_i\, \alpha(\vartheta_i), \tag{4.20}$$

where $\alpha(\vartheta_i)$ is defined by Eq. (4.17). We reduce the power carried by the ray by an amount P_a and then compare its remaining power $P_r = P_i - P_a$ with a threshold value. For example, we might terminate the ray and launch a new ray if $P_r < P_o \times 10^{-6}$, where P_o is the original power carried by the ray before its first reflection. Otherwise, the ray continues at its reduced strength. This "slow death" approach can also be used in diffuse-specular gray analysis, but it must be used in non-diffuse non-gray analysis. The general step-by-step procedure for computing band-averaged spectral radiation distribution factors is presented in Section 4.6.

While prior knowledge of the value of absorptivity α was available in Chapter 3, this is not the case in the current chapter. Therefore, we must first compute the directional absorptivity $\alpha(\vartheta_i)$. This requires that a ray trace be performed in which an appropriately large number M of reflections are presumed to occur from the point of incidence.

Referring to Figure 4.7, a random direction $\boldsymbol{v}_r$ is assigned to each of the M reflections using Kowsary's method described in Step 8 of Section 3.3. The direction $\boldsymbol{v}_v$ is obtained using Eq. (4.16) with $x_v = \sin\vartheta_v$, $y_v = 0$ (since $\vartheta_v = \vartheta_i$ and $\varphi_v = 0$), and $z_v = \cos\vartheta_v$. The angle γ is then computed using Eq. (4.14), after which Eqs. (4.10) and (4.11) are used to compute the bidirectional reflectivity, $\rho''(\vartheta_i, \vartheta_r, \varphi_r)$. Finally, the directional absorptivity is computed using Eq. (4.17).

The method may be effectively demonstrated by simulating measurement of the *BRDF* using the apparatus illustrated schematically in Figure 4.3. In the simulation, we illuminate the Z302-coated coupon with a collimated 2.0-mm diameter circular beam of power P_s incident at 45°. The beam is composed of a bundle of $N = 1000$ randomly spaced parallel rays, each carrying a power of $P_i = 10.0$ mW. Upon incidence to the Z302-coated coupon, each ray is split into $M = 100\,000$ raylets that are reflected into the hemispherical space above their point of incidence. Note that the illuminated spot in Figure 4.3 has a finite area. In the simulation we set the distance R in Figure 4.3 to 100 mm to ensure that the detector lies in the far field of the intersection of the beam with the coupon.

Figure 4.11 shows the simulated power measured by the detector as it is displaced at constant azimuth angle $\varphi_d = 0$ over a range of zenith angles $15 \leq \vartheta_d \leq 75°$. The power measured by the detector depends on the number of rays incident to it, which, in turn, depends on the solid angle subtended by the detector at the center of the illuminated spot on the coupon. Reference to Figures 4.9 and 4.10 suggests that about 90% of the incident power, about 9 W in this case, is expected to be absorbed, while the remaining 1 W or so is reflected into the hemispherical space above the coupon.

It now remains to convert the power distribution into *BRDF*. The power incident to the detector for a given incidence angle ϑ_i and corresponding solid angle Ω_i due to N incident rays and M reflected rays per incident ray is

$$P_d = \langle i_d \rangle A_d \Omega_d, \tag{4.21}$$

where

$$\langle i_d \rangle = i_i \langle BRDF_d \rangle \cos\vartheta_i\, \Omega_i. \tag{4.22}$$

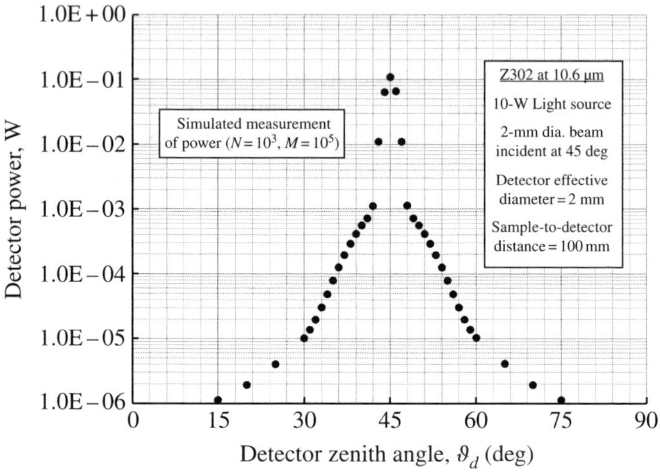

Figure 4.11 Power measured by the detector using the apparatus of Figure 4.3 [4].

In Eq. (4.22), $i_i = P_s / A_s \Omega_s$ and $\langle BRDF_d \rangle$ is the mean value of the *BRDF* corresponding to the detector location and effective area:

$$\langle BRDF_d \rangle \equiv \frac{1}{M_d} \sum_{M_d} BRDF_{m_d}, \tag{4.23}$$

where M_d is the number of reflected rays collected by the detector. Recalling that the apparatus has been designed so that $A_s\Omega_s = A_d\Omega_d = A\Omega$, we combine Eqs. (4.21) and (4.22) to obtain

$$\langle BRDF_d \rangle = \frac{P_d/P_s}{\Omega \cos \vartheta_i} = \frac{P_d/P_s}{(\pi r/R)^2 \cos \vartheta_i}. \tag{4.24}$$

With the introduction of numerical values from the current example into Eq. (4.24), we have $\langle BRDF_d \rangle \cong 450\ P_d$. Thus, the peak value in Figure 4.11, where $P_d = 0.109$ W, corresponds to a *BRDF* of 49 sr^{-1}, which is approximately one-half the "true" value calculated directly using the four-component model. This apparent discrepancy may be attributed to several factors, but it is mainly due to the finite size of the detector, whose angular extent is 1.145°, and to the local rate of change of *BRDF* with detector zenith angle near 45°. The apparatus measures the mean value of *BRDF* over the effective detector area. Therefore, results obtained near the reflection peak significantly underestimate the real

BRDF values. Other factors contributing to the low value of measured *BRDF* are the finite size of the illuminating beam, whose diameter is 2.0 mm, and the limited distance from the coupon to the detector, 100 mm. The inherent inaccuracy of the numerical model itself ($N = 1000$ incident rays and $M = 100\,000$ reflected rays) must also be considered, although this can be minimized by increasing N and M until the results obtained are no longer sensitive to their values. When the source and detector diameters are decreased from 2.0 to 1.0 mm and the number of rays traced is increased from $N = 1000$ to $N = 10\,000$, the value obtained by the simulation for *BRDF* at $\varphi_d = 45°$ increases from 49 to 81 sr^{-1}.

Modeling the apparatus using the MCRT method permits a parametric study to be carried out to determine the optimum values of P_s, r, and R. For example, inspection of Figure 4.11 reveals that if a single detector is to be used at all reflection angles, it must have a dynamic range covering six orders of magnitude and a sensitivity extending down into the low microwatt range. Furthermore, increasing the source power to obtain a higher detector power at large incidence angles might produce intolerably high heat fluxes on the coupon. Similarly, decreasing the radius r to increase the accuracy of the *BRDF* measurement near the peak also increases the local heat flux on the sample. The MCRT method is clearly a tremendously valuable tool for experimental design and optimization.

4.4 Second Example: A Highly Reflective Surface Whose Reflectivity is Strongly Diffuse

We next consider a practical application whose accurate simulation requires a bidirectional spectral reflection model. The **integrating sphere** [5] often plays a central role in radiometric instrument calibration, surface optical behavior measurement, and radiant source characterization [6, 7]. The purpose of an integrating sphere is to convert a collimated beam of monochromatic light, such as might be provided by a laser source, into a larger, weaker source of diffuse light at the same wavelength. The essential property of the integrating sphere is its ability to produce a Lambertian source of monochromatic radiation due to multiple scattering from its interior walls. This requirement will be satisfied exactly for a completely enclosed spherical cavity, even when the wall coating is not itself a perfectly diffuse reflector. However, a practical integrating sphere must be fitted with ports that allow the illuminating beam to enter and the instrument under calibration to view the interior wall. The ports inevitably allow some of the entering radiation to escape before being completely diffused by reflections, thereby compromising the desired effect. In practice the ports are made as small as possible compared to the diameter of the sphere, and the interior walls are treated with a highly reflective, highly diffuse coating.

The author and his coworkers [8] have investigated the departure from ideal behavior of a practical integrating sphere, with emphasis on the influence of directionality. The results of that investigation are offered here as an example of an application in which the diffuse gray assumption may be inadequate. We consider an application in which a relatively small integrating sphere is to be used on-orbit to calibrate a radiometer against a reference radiometer by having both instruments observe the same sector of the interior wall through two separate ports. Because the calibrations will be repeated over a period of up to several years, it is important to know how the calibration factor might be expected to vary with aging of the wall coating. It is further interesting to know how the degree of directionality of reflections from the walls might influence its performance.

We use the MCRT method to simulate illumination of the interior by a quasi-monochromatic light source at a wavelength, 0.9 μm, for which the *BRF* model developed below may be considered valid. For purposes of the current investigation the diameter of the port through which the narrow light beam is admitted may be considered sufficiently small compared to the diameters of the two viewing ports to neglect its presence in the ray trace. The hypothetical experimental arrangement is illustrated in Figure 4.12.

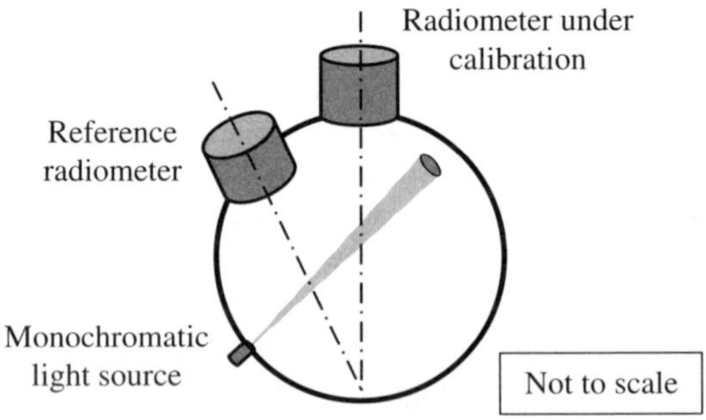

Figure 4.12 An integrating sphere configured to compare the responses of two radiometers to the same quasi-diffuse target.

The light source is positioned and aimed so that its spot of initial incidence on the sphere wall lies outside the common field-of-view of the two radiometers. Radiation escaping through the two radiometer ports is assumed to be completely lost, while radiation incident to the sphere walls is reflected to a degree and with a directionality governed by the Spectralon® *BRF* model developed in this section. The diameter of the integrating sphere is 100 mm, the effective diameter of the radiometer under calibration is 33.8 mm, and the effective diameter of the reference radiometer is 32.0 mm. The spread half-angle of the illuminating source is 10°, and the length of the beam is the diameter of the sphere. It is clear that, while the two instruments can be positioned so that they view essentially the same patch of wall, any directionality of the wall reflectivity will lead to differences in measured flux.

What is required is a bidirectional reflectivity model for use in the MCRT environment that produces reflected rays whose directional distribution is random, as in the case of diffuse reflection, but whose power in a prescribed direction is

$$P_r = P_i \rho BRF. \tag{4.25}$$

In Eq. (4.25), ρ is the wall bi-hemispherical reflectivity determined as described in Section 4.3, Eq. (4.7). The results presented here are limited to the visible and near-infrared wavelength interval ($0.4 \leq \lambda \leq 0.9\ \mu\text{m}$), so with this understanding the subscript λ is suppressed in the following development.

As was the case for Z302, the form of the *BRF* model for Spectralon is suggested by the data upon which it is based. The approach is somewhat different in this case because the bidirectionality of Spectralon is rather weak. We begin by identifying three modeling parameters: a pseudo-specular component of reflectivity ρ^s, a pseudo-diffuse component of reflectivity ρ^d, and the standard deviation σ of the angle between a ray reflected in the direction of the observed peak reflection and a diffusely reflected ray. The basic form of the bidirectional reflectivity model is

$$BRF = \rho^s + (\rho^d - \rho^s)\left(1 - e^{\frac{-\delta^2}{2\sigma^2}}\right), \tag{4.26}$$

where

$$\delta = \cos^{-1}[\boldsymbol{v}_p \cdot \boldsymbol{v}_d]. \tag{4.27}$$

In Eq. (4.27), $\boldsymbol{v}_p$ is the unit vector in the direction of the peak and $\boldsymbol{v}_d$ is the unit vector in the direction of the diffuse reflection.

We begin by deriving empirical relations for the variation of the three model parameters with incident zenith angle, ϑ_i. In this example the resulting curve-fitting relationships are based on data [9] for a single wavelength, 0.9 μm. The wavelength dependence of *BRF* for Spectralon between 0.4 and 0.9 μm is sufficiently weak that the results reported here may be assumed to apply over that wavelength interval. The relationships thus obtained are

$$\rho^s = 0.089\,(12 + 0.06e^{0.07\vartheta_i}), \tag{4.28}$$

$$\rho^d = 0.92 - 0.07\,\cos\left(\frac{4\pi\vartheta_i}{180}\right), \tag{4.29}$$

and

$$\sigma = 0.27(1.0 + e^{0.09\,\vartheta_i}). \tag{4.30}$$

The units of ϑ_i are degrees in Eqs. (4.28)–(4.30).

These correlations lead to a directional reflection model for Spectralon that it is convenient to use in the MCRT environment but may be inadequate for applications where even higher accuracy is required. However, the model is deemed to be more than adequate for realistically revealing the effect of weakly directional reflections on the performance of a practical integrating sphere with multiple wall openings. The technique with appropriate refinements and perhaps involving reference to a broader population of experimental data would no doubt lead to increased accuracy when optimized to a specific application.

Figure 4.13 compares values of the correlating parameters returned by Eqs. (4.28)–(4.30) with discrete values corresponding to specific incident angles 0, 30, 45, and 60°. The symbols in these curves represent values obtained by adjusting the appropriate parameter to obtain a best agreement between the measured data (solid symbols in Figure 4.14) and the Model A result (open symbols in Figure 4.14). Extrapolations of the curves beyond the range of available data in Figure 4.13 are informed by theory and justified by the extent of agreement between measured and modeled *BRF* values shown in Figure 4.14.

With one exception, corresponding to the standard deviation at $\vartheta_\mathrm{i} = 30°$, the functions used to compute the values of ρ^s, ρ^d, and σ in the MCRT model provide excellent fits to the data. In this exceptional case the one errant point was simply ignored since there is no way either to explain it on the basis of theory or to include it in a reasonably simple function. This is perhaps a good occasion to emphasize that the justification for these functions is that they can be used in fast

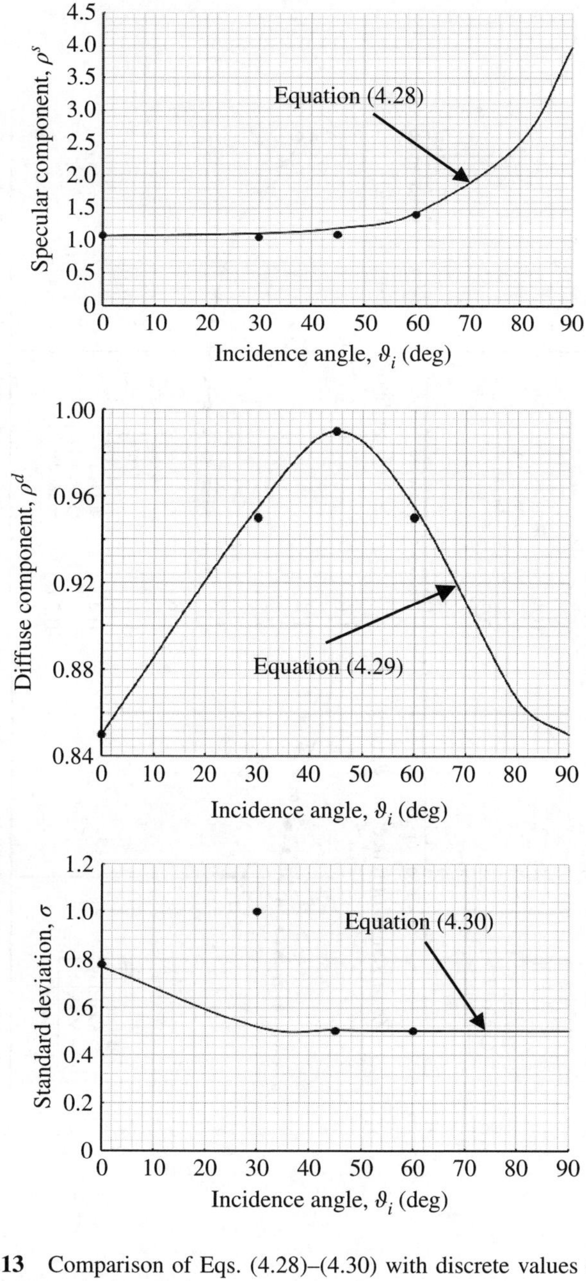

Figure 4.13 Comparison of Eqs. (4.28)–(4.30) with discrete values of fitting parameters [8].

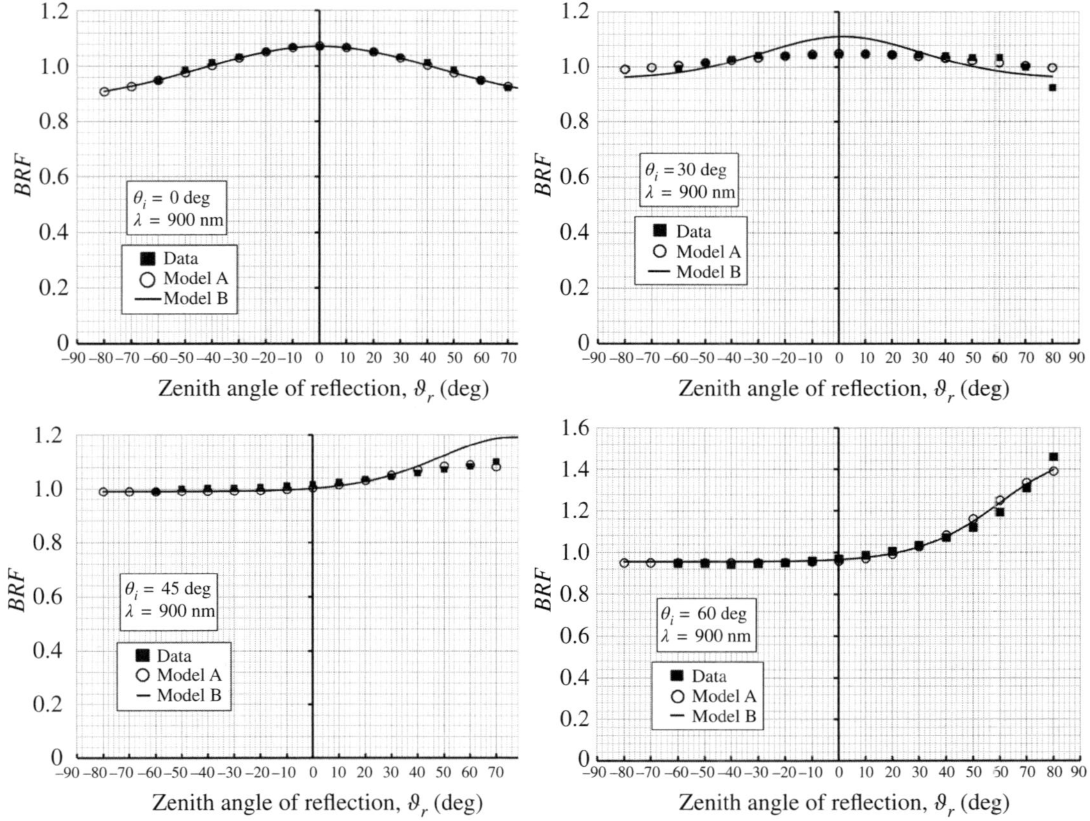

Figure 4.14 Comparison of *BRF* data from Walker [9] with two models [8].

algorithms to model millions of directional reflections without the need to interpolate in tables. The consequence of the decision to ignore this single data point is the relatively poor agreement between the measured data (closed symbols) and the results returned by the MCRT model (curve labeled Model B) in Figure 4.14.

Further fitting is required to account for **off-specular peaking**, a phenomenon in which the peak in the directional reflectance distribution occurs at a larger-than-specular zenith angle (see Ref. [1], p. 127). The Spectralon data used in the current effort are also the basis for correlating the peak zenith angle with the incident zenith angle using the **sigmoidal function**

$$\vartheta_{peak} = \frac{\pi/2}{1 + e^{[(a\pi/2 - 2\vartheta_i)/b]}}, \tag{4.31}$$

where $a = 0.9$ is a "shifting" factor and $b = \pi/36$ is a "stretching" factor. The angles in Eq. (4.31) are in radians.

Once expressions are obtained for ρ^s, ρ^d, σ, and ϑ_{peak} in terms of ϑ_i for a given wavelength interval (the visible and near infrared in this case), they may be introduced into the statistical model embodied in Eqs. (4.26) and (4.27).

Figure 4.14 compares the measured *BRF* values from the literature [8] (closed symbols) with those obtained using the individually adjusted values of ρ^s, ρ^d, σ, and ϑ_{peak} (open symbols, Model A) and the values obtained using Eqs. (4.28)–(4.31) (curve, Model B). The agreement between the original *BRF* data and the corresponding model results could probably be improved, for example by using higher-order correlations or a more sophisticated model. Still, the agreement obtained here preserves the trends and the general range of the data, so is judged to be adequate for meeting the goal of establishing the sensitivity of practical integrating sphere performance to weak directionality of the wall coating.

Two reflection models are used in a Monte Carlo ray-trace to describe the distribution of the radiation from the source to the two radiometer ports: (i) a diffuse model and (ii) the Spectralon *BRF* model developed here. For the diffuse model, the energy of a ray diminishes by a factor of $\rho = 1 - \alpha$ for each reflection. The direction of each reflection is determined randomly following the usual rules of a Monte Carlo ray-trace. A counter is incremented when the ray enters one of the two ports and then a new ray is launched. The same procedure is employed for the Spectralon *BRF* model with the exception that, in addition to multiplying the power of each reflected ray by $\rho = 1 - \alpha$, the power is also multiplied by the value of *BRF* corresponding to the angle of incidence, ϑ_i.

In the absence of the two radiometer access ports the wall radiance would be uniform and have the same value for both reflection models. Then, in this special case, the ratio of the power incident to a wall sector the size of the opening provided for the radiometer under calibration to the power incident to a wall sector the size of the opening provided by the reference radiometer would simply be their area ratio $(16.9/16.0)^2 \approx 1.115$. Figure 4.15 shows the variation with wall absorptivity α of the actual radiometer power ratio for two cases: (i) purely diffuse reflections and (ii) directional reflections governed by the Spectralon *BRF* model.

Figure 4.15 responds directly to the question, What is the sensitivity of the radiometer power ratio to the degradation over time of the wall coating? Of course, we would like to see a complete insensitivity of the radiometer power ratio to wall absorptivity, which would be represented in Figure 4.15 by a horizontal line. Instead we see a linear relationship in which a 1% change in the wall absorptivity translates into a more than 2% change in the calibration factor. The range of radiometer power ratios, and thus of the calibration factor, for both reflection models exceeds the "ideal" value based on area ratio alone (=1.115) by about 50%. One interesting and unexpected advantage of a slightly directional coating over a purely diffuse coating, at least for the geometry and wall coating currently

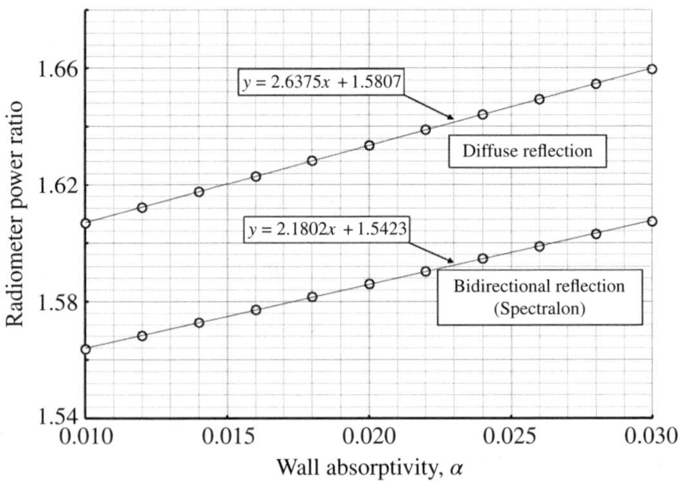

Figure 4.15 Ratio of power distributed to the two radiometers as a function of wall absorptivity (note that the vertical axis does not start at zero). Source [8, reprinted with permission].

under consideration, is the nearly 20% reduction in the sensitivity of the calibration factor, from 2.6375 (diffuse) to 2.1802 (bidirectional).

It is appropriate to speculate about the generality of this result in view of the large variability of geometry possible in such applications. Sintered PTFE, of which Spectralon is a commercial example, represents a practical limit on diffuse behavior available for integrating sphere coatings. Any increase in wall opening fraction must lead to fewer reflections and, thus, to an increase in the influence of wall absorptivity. In fact, in the limit as the geometry approaches that of a plane wall with a single reflection, a 1% change in the surface absorptivity translates into a 1% change in the calibration factor. This effect alone would explain the decrease in slope of the line in Figure 4.15 when reflectivity is changed from diffuse to slightly directional.

It may be concluded that, for the case of an integrating sphere with two openings, the ratio of the power escaping through the two openings varies linearly with the absorptivity of the wall. For the apparatus considered in the current contribution, the fractional change in the value of the calibration factor, defined as the ratio of the power exiting the two ports, is more than twice that of the wall absorptivity. Finally, even when the departure of the wall reflectivity from Lambertian is relatively minor, as in the case of Spectralon, a significant reduction in the sensitivity of the calibration factor to the wall absorptivity is associated with directional reflections.

4.5 The Band-Averaged Spectral Radiation Distribution Factor

The case studies presented in Sections 4.3 and 4.4 exemplify direct application of the MCRT method without recourse to radiation distribution factors, which were not needed to accomplish the stated goals. Furthermore, they involve situations for which the wavelength interval of interest is sufficiently narrow that the surface models used are, to an acceptable approximation, independent of wavelength. However, in some cases radiation distribution factors are required, as explained in Chapter 3. The radiation distribution factor introduced and used in Chapter 3 for gray surfaces had two subscripts, *i* and *j*, the indices of the emitting and absorbing surface. For the case of spectral radiation it is necessary to add a third subscript, *k*, representing the wavelength interval $\Delta\lambda_k$ in which the distribution factor applies. We define the **band-averaged spectral radiation distribution factor** D_{ijk} as the fraction of power emitted in wavelength interval $\Delta\lambda_k$ by surface element i that is absorbed by surface element j, both directly and due to all possible reflections within the enclosure.

Estimation of the band-averaged spectral radiation distribution factor matrix assumes the availability of a dense data set ultimately based on extensive laboratory measurements. Imagine a bookshelf in a virtual thermophysical properties library bearing the label "Bidirectional Spectral Reflectivity." Upon perusal of this bookshelf we might find books with titles "Z302," "Spectralon," "Gold Black," and other optical coatings. When we take down one of these books and open to its table of contents; we find chapter titles such as "Wavelength Interval Between 0.01 and 0.10 μm," and "Wavelength Interval Between 0.10 and 1.00 μm," and so forth. Then when we flip through Chapter 1 we notice page headings "Angle of Incidence = 5°," "Angle of Incidence = 10°," and, on the last page, "Angle of Incidence = 85°." Finally, when we scan one of these pages from top to bottom we find on the first line "Angle of Reflectance = 5°, *BRDF* = 44.21," and on the second line "Angle of Reflectance = 10°, *BRDF* 38.45," and so forth. Upon plotting the data found in one of these books, we recognize that the spacing between successive wavelengths, angles of incidence, and angles of reflectance is sufficiently small to allow accurate linear interpolation between tabulated values.

Unfortunately, the bookshelf described above is not yet available, at least not to the public, although certain government agencies and private laboratories no doubt possess some of these books. One can hope that eventually an on-line service will exist that makes such data more widely available. In the meantime it is often necessary to undertake costly campaigns to measure and archive the bidirectional spectral reflectivity data for a given application. Assuming that such data are available, the following steps describe their use in computing the elements of the band-averaged spectral radiation distribution factor matrix.

***Step 1*: Determine the wavelength interval index k of the emitted ray.**

The wavelength interval of the emitted ray will depend on the temperature of the emitting surface element. This requires that the element temperature be known. (An iterative approach, such as the one presented in Section 4.7, must be used when the element temperature is unknown.) The probability that a ray will be emitted at a specified wavelength λ is

$$P(\lambda) = \frac{\int_{\lambda=0}^{\lambda} i_{b\lambda}(\lambda', T_i)d\lambda'}{\int_{\lambda=0}^{\infty} i_{b\lambda}(\lambda', T_i)d\lambda'}, \quad 1 \leq i \leq n. \tag{4.32}$$

Then, in principle, we draw the next available random number R_λ, equate it to $P(\lambda)$, and "solve" Eq. (4.32) for the corresponding wavelength λ.

The denominator of Eq. (4.32) is, of course, $\sigma T_i^4/\pi$, so we are left with the chore of solving the integral equation

$$\int_{\lambda=0}^{\lambda} i_{b\lambda}(\lambda', T_i)d\lambda' = P(\lambda)\sigma\frac{T_i^4}{\pi} \tag{4.33}$$

millions of times during a typical ray trace. Therefore, it is essential to define a suitable approximation for the integral as a function of λ that can be solved explicitly for λ as a function of $P(\lambda)$. Figure 4.16 compares the function on the right-hand side of Eq. (4.32) with the approximating function

$$P(\lambda) \approx 1 - e^{-a(\lambda-\lambda_{\min})}, \quad \lambda > \lambda_{\min}, \tag{4.34}$$

when $T_i = 800$ K. The approximating function requires only two adjustable constants, a and $\lambda_{\min}$, and is readily solved explicitly for λ as a function of $p(\lambda)$,

$$\lambda = \lambda_{\min} - \frac{1}{a}\ln(1 - P), \quad P < 1. \tag{4.35}$$

Equation (4.35) is applicable only over the limited wavelength range $\lambda_{\min} \leq \lambda \leq \lambda_{\max}$. As a practical matter, the random number generator will never return a value of exactly unity, but logic still must be provided to

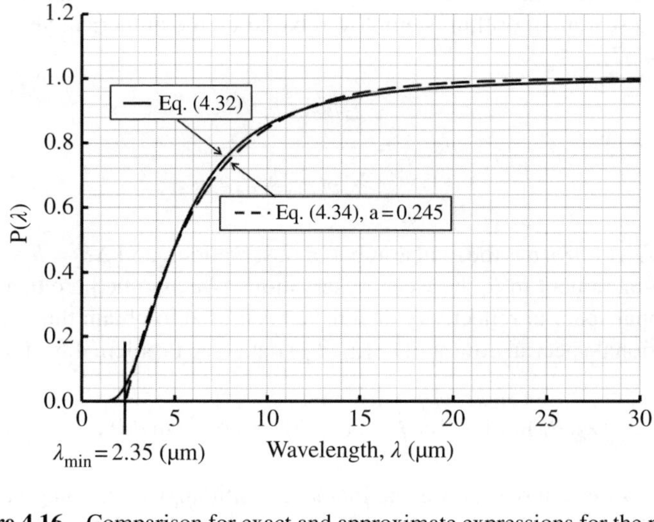

Figure 4.16 Comparison for exact and approximate expressions for the probability $P(\lambda)$ that a ray will be emitted at wavelength λ when $T_i = 800$ K.

limit the maximum wavelength to a range covered by the surface emissivity model. The values of a and λ_{min} must be selected with this in mind.

Once the wavelength of emission λ has been determined, the corresponding value of the index k can be assigned depending on the number of subintervals into which the interval $\lambda_{max} - \lambda_{min}$ has been divided.

***Step 2*: Determine the direction cosines of the emitted ray.**

Most surfaces of practical engineering interest are isotropic; that is, the directional spectral emissivity is independent of the azimuthal angle φ_E. It can be shown (see Problem 4.3) that the *relative* probability that a ray will be emitted at a zenith angle ϑ_E is

$$p(\lambda_k, \vartheta_E) = \varepsilon(\lambda_k, \vartheta_E) \cos \vartheta_E \sin \vartheta_E, \tag{4.36}$$

and that $p(\lambda_k, \vartheta_E)$ is related to a random number R_{ϑ_E}, which is uniformly distributed between zero and unity, according to

$$R_{\vartheta_E} = \int_0^{\vartheta_E} p(\lambda_k, \vartheta')d\vartheta' \Big/ \int_0^{\pi/2} p(\lambda_k, \vartheta')d\vartheta'. \tag{4.37}$$

We draw the next available random number, R_{ϑ_E}, and use it to compute ϑ_E. In the case of a diffuse surface, Eq. (4.37) yields an explicit expression for ϑ_E, the now familiar $\vartheta_E = \sin^{-1}\left[\sqrt{R_{\vartheta_E}}\right]$ (see Problem 4.4). However, in situations where measured directional emissivity data must be used, a search algorithm is employed based on the numerical version of Eq. (4.37),

$$R_{\vartheta_{E,i}} = \frac{\sum_1^i p(\lambda_k, \vartheta_i)\Delta\vartheta_i}{\sum_1^n p(\lambda_k, \vartheta_i)\Delta\vartheta_i}. \tag{4.38}$$

In Eq. (4.38), n is the total number of observations, and $\Delta\vartheta_i = \vartheta_i - \vartheta_{i-1}$ is the increment in ϑ_E between observations. We assume here that bidirectional spectral reflectivity data have been used to obtain the required directional spectral emissivity $\varepsilon(\lambda_k, \vartheta_E)$ data (see Problem 4.5); that is,

$$\varepsilon(\lambda_k, \vartheta_E) = 1 - 2\pi \int_0^{\frac{\pi}{2}} \rho(\lambda_k, \vartheta_E, \vartheta) \cos \vartheta \sin \vartheta d\vartheta, \tag{4.39}$$

where ϑ_E is a surrogate for the incident zenith angle, and integration is over the reflected zenith angle. Of course, numerical evaluation of the integral in Eq. (4.39) would be required. Once $\vartheta_{E,i}$ is so determined and it is recognized that $\varphi_E = 2\pi R_{\varphi_E}$ (the isotropic assumption), Eq. (1.34) is used to compute the direction cosines L_E, M_E, and N_E of the emitted ray.

***Step 3*: Determine the index *j* of the surface element containing the point of intersection of the emitted ray with a surface of the enclosure.**

The procedure is the subject of Chapter 1 and is demonstrated in Chapter 3. We use an indexing algorithm as described in Section 1.6 to determine the index of the surface element containing the point of intersection.

***Step 4*: Compute the directional spectral absorptivity of the surface of incidence *j*.**

The directional spectral absorptivity at a given wavelength is equal to the directional spectral emissivity (Kirchhoff's law); that is,

$$\alpha_j(\lambda_k, \vartheta_{I,j}) = \varepsilon(\lambda_k, \vartheta_{I,j}). \tag{4.40}$$

In Eq. (4.40), $\vartheta_{I,j}$ is the incident zenith angle,

$$\vartheta_{I,j} = \cos^{-1}[-\boldsymbol{v_I} \cdot \boldsymbol{n}], \tag{4.41}$$

where $\boldsymbol{v_I} = L_E\,\boldsymbol{i} + M_E\,\boldsymbol{j} + N_E\,\boldsymbol{k}$ is the unit vector describing the direction of the incident ray and $\boldsymbol{n}$ is the local unit normal vector at the point of intersection of the ray with the surface. Evaluation of the absorptivity then requires that wavelength information in the form of the index k be carried by the ray. The directional spectral absorptivity $\alpha_j(\lambda_k, \vartheta_{I,j})$ is then obtained by first applying Eq. (4.39) with the "*E*" subscript replaced by "*I,j*" to obtain $\varepsilon(\lambda_k, \vartheta_{I,j})$, and then applying Kirchhoff's Law, Eq. (4.40).

***Step 5*: Compute the fraction of the power carried by the incident ray that is absorbed in surface element j.**

Once the directional spectral absorptivity has been obtained, a fraction $\alpha_j(\lambda_k, \vartheta_{I,j})$ of the power carried by the incident ray is left behind and a reduced-power ray is reflected. The power absorbed by surface element j is incremented by an amount $\alpha_j(\lambda_k, \vartheta_{I,j})P_{ijk}$, where

$$\begin{aligned} P_{ijk} &= \varepsilon_i(\lambda_k, \vartheta_{E,i}) \int_{\Delta\lambda_k} i_{b\lambda}(\lambda, T_i)\,d\lambda A_i \cos\vartheta_{E,i}\Omega_{ij} \\ &= \varepsilon_i(\lambda_k, \vartheta_{E,i}) i_b(\Delta\lambda_k, T_i) A_i \cos\vartheta_{E,i}\Omega_{ij} \end{aligned} \tag{4.42}$$

is the power carried by the incident ray. In Eq. (4.42), $\varepsilon_i(\lambda_k, \vartheta_{E,i})$ is the directional spectral emissivity of surface element i in wavelength interval k corresponding to zenith angle $\vartheta_{E,i}$, $\Delta\lambda_k = \lambda_k - \lambda_{k-1}$ is the width of wavelength interval k, and $\Omega_{ij} = A_j \cos \vartheta_{I,j}/t_{ij}$ is the solid angle subtended by surface element j at surface element i. The power of the incident ray is reduced by a factor of $\rho_j(\lambda_k, \vartheta_{I,j}) = 1 - \alpha_j(\lambda_k, \vartheta_{I,j})$ and the ray is then allowed to reflect, provided that it still carries sufficient power. The strength of the ray after each reflection is compared with a preset threshold value to determine when it should be terminated.

The process whereby the strength of the ray decreases with each reflection until it is finally extinguished is referred to as the "slow death" approach to the MCRT method. This is in contrast with the "sudden death" approach followed in Chapter 3 in which the ray maintains its original strength until all of its power is suddenly absorbed. Note that both the sudden death and the slow death versions of the MCRT method work equally well in diffuse gray enclosures, but only the slow death approach can be applied in non-diffuse, non-gray enclosures.

***Step 6*: Determine the direction cosines of the reflected ray.**

Assuming that a reflection occurred as a result of Step 5, the next step is to determine the direction cosines of the reflected ray. As in the example problem of Section 4.4, we compute the direction of reflection using Kowsary's method described in Step 8 of Section 3.3. The assumption of a diffuse directional distribution of reflection is justified in this step because the power of each reflected ray is not determined until later, when we apply the bidirectional reflectivity in Step 8.

***Step 7*: Determine the index of the surface element containing the point of intersection of the reflected ray with a surface of the enclosure.**

Once again, the procedure for accomplishing this step is given in Chapter 3.

***Step 8*: Determine the power carried by the reflected ray.**

We imagine that, upon reflection, the incident ray splits into a large number of reflected rays, with the power of each individual reflected ray given by

$$P_{R,jk} = P_{ijk}\rho_j(\lambda_k, \vartheta_{I,j}, \vartheta_{R,j}, \varphi_{R,j}) \cos \vartheta_{R,j}\Omega_{R,j}, \tag{4.43}$$

where $\rho_j(\lambda_k, \vartheta_{I,j}, \vartheta_{R,j}, \varphi_{R,j})$ is the bidirectional reflectivity of surface element j, $(\vartheta_{R,j}, \varphi_{R,j})$ is the direction of the reflection with respect to the local coordinate system, and $\Omega_{R,j}$ is the solid angle containing the reflected ray. This latter quantity is the projected area of the surface element intersected by the reflected ray divided by the distance to this surface from surface *j*. A version of Eq. (4.41) is used to find the required projected area. Finally, from the definition of the radiation distribution factor we have

$$D_{ijk} \approx \frac{Q_{ijk}}{Q_{ik}}, \tag{4.44}$$

where Q_{ik} is the power in wavelength interval $\Delta\lambda_k$ emitted by surface element *i* and Q_{ijk} is the power accumulated by surface element *j* in wavelength interval *k* due to energy bundles emitted from surface element *i* in that wavelength interval.

4.6 Use of the Band-Averaged Spectral Radiation Distribution Factor for the Case of Specified Surface Temperatures

Consider the case where the temperatures of all *n* surface elements are specified, and where the surface heat flux on one or more surfaces is sought. The power (W) absorbed by surface element *i* may be computed directly from the definition of the band-averaged spectral radiation distribution factor; that is,

$$Q_{i,a} = \sum_{k=1}^{K} \sum_{j=1}^{n} \varepsilon_{jk} e_b(\Delta\lambda_k, T_j) A_j D_{jik}, \quad 1 \le i \le n, \tag{4.45}$$

where ε_{jk} is the hemispherical band-averaged spectral emissivity of surface element j,

$$\begin{aligned} \varepsilon_{jk} = \varepsilon_j(\lambda_k) &\equiv \frac{1}{\pi} \int_{2\pi} \varepsilon_j(\lambda_{k,} \vartheta, \varphi) \cos \vartheta d\Omega \\ &= 2 \int_0^{\pi/2} \varepsilon_j(\lambda_{k,} \vartheta) \cos \vartheta \sin \vartheta d\vartheta, \end{aligned} \tag{4.46}$$

and $e_b(\Delta\lambda_k, T_j)$ is the blackbody emissive power contained in the wavelength interval $\Delta\lambda_k = \lambda_k - \lambda_{k-1}$,

$$\mathrm{e_b}(\Delta\lambda_\mathrm{k}, \mathrm{T_j}) = \pi \int_{\lambda_{\mathrm{k}-1}}^{\lambda_\mathrm{k}} \mathrm{i_b}(\lambda, \mathrm{T_j}) d\lambda. \tag{4.47}$$

Reciprocity holds for the band-averaged spectral radiation distribution factor; that is,

$$\varepsilon_{ik} A_i D_{ijk} = \varepsilon_{jk} A_j D_{jik}. \tag{4.48}$$

Then the flux (W m^{-2}) absorbed by element i may be found by introducing Eq. (4.48) into Eq. (4.45) and dividing through by A_i, yielding

$$\frac{Q_{i,a}}{A_i} = q_{i,a} = \sum_{k=1}^{K} \sum_{j=1}^{n} \varepsilon_{ik} e_b(\Delta\lambda_k, T_j) D_{ijk}, \quad 1 \leq i \leq n. \tag{4.49}$$

Finally, the *net* heat flux *from* element i is given by

$$q_i \equiv q_{i,e} - q_{i,a} = \sum_{k=1}^{K} \sum_{j=1}^{n} \varepsilon_{ik} e_b(\Delta\lambda_k, T_j)(\delta_{ijk} - D_{ijk}), \quad 1 \leq i \leq n, \tag{4.50}$$

where now the Kronecker delta is defined

$$\delta_{ijk} \equiv \begin{cases} 1, & i = j \\ 0, & i \neq j \end{cases}, \quad \text{all } k. \tag{4.51}$$

4.7 Use of the Band-Averaged Spectral Radiation Distribution Factor for the Case of One or More Specified Surface Net Heat Fluxes

If one or more surface net heat fluxes is specified, we are faced with the complication that we cannot solve Eq. (4.50) explicitly for T_j. Consider the case where N of the n surface elements have specified surface net heat fluxes, and the remaining $n - N$ surface elements have specified temperatures. We begin by splitting the right-hand side of Eq. (4.50), written for only the first N surfaces, into two summations,

$$q_i \equiv q_{i,e} - q_{i,a} = \sum_{k=1}^{K} \left[\sum_{j=1}^{N} \varepsilon_{ik} e_b(\Delta\lambda_k, T_j)(\delta_{ijk} - D_{ijk}) - \sum_{j=N+1}^{n} \varepsilon_{ik} e_b(\Delta\lambda_k, T_j) D_{ijk} \right], \quad 1 \leq i \leq N. \tag{4.52}$$

In Eq. (4.52), the first N terms in the brackets on the right-hand side involve the N unknown temperatures, while the remaining $n - N$ terms involve only known quantities. It is convenient to rewrite Eq. (4.52)

$$\sum_{k=1}^{K}\sum_{j=1}^{N}\varepsilon_{ik}e_b(\Delta\lambda_k,T_j)(\delta_{ijk}-D_{ijk})=q_i+Q_i,\qquad 1\le i\le N,\qquad (4.53)$$

where

$$Q_i=\sum_{k=1}^{K}\sum_{j=N+1}^{n}\varepsilon_{ik}e_b(\Delta\lambda_k,T_j)D_{ijk},\qquad 1\le i\le N.\qquad (4.54)$$

Thus, the right-hand side of Eq. (4.53) is a known vector of length N, so Eq. (4.53) itself represents N equations in N unknown surface temperatures. Note that these equations cannot be solved explicitly, because $e_b(\Delta\lambda_k, T_j)$ is an intrinsic function of T_j. However, once Eq. (4.53) has somehow been solved for the N unknown surface temperatures, we do have available an explicit expression for the $n-N$ unknown surface net heat fluxes,

$$q_i=\sum_{k=1}^{K}\sum_{j=1}^{n}\varepsilon_{ik}e_b(\Delta\lambda_k,T_j)(\delta_{ijk}-D_{ijk}),\qquad N+1\le i\le n.\qquad (4.55)$$

Solution of Eq. (4.53) is not straightforward when more than one surface has a specified net heat flux. However, in the special but common situation where only a single surface has a specified net heat flux, Eq. (4.53) may be rewritten

$$f(T_1)=\sum_{k=1}^{K}\varepsilon_{1k}e_b(\Delta\lambda_k,T_1)(1-D_{11k})-q_1$$
$$-\sum_{k=1}^{K}\sum_{j=2}^{n}\varepsilon_{1k}e_b(\Delta\lambda_k,T_j)D_{1jk}=0.\qquad (4.56)$$

We now seek an algorithm capable of solving this transcendental equation for the root T_1. The Newton–Raphson iteration is a possibility but it requires numerical estimates of $f'(T_1)$, the derivative of $f(T_1)$, at each step. A far simpler approach is the half-interval search, illustrated in Figure 4.17. The half-interval search for a root involves systematically incrementing T_1 through a sequence of equally spaced values such that $T_1^{m+1}=T_1^m+\Delta T_1$, where ΔT_1 is a constant whose value determines the accuracy of the solution. For each value of T_1, Eq. (4.56) is used to

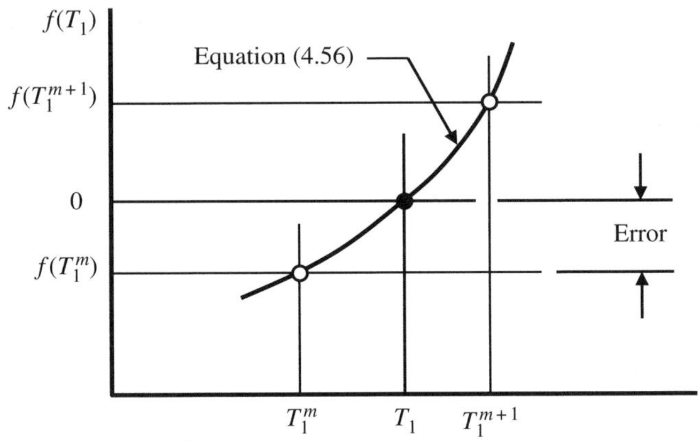

Figure 4.17 Illustration of the half-interval search method for finding the roots of a transcendental equation.

compute the corresponding value of $f(T_1)$. At each step the sign of the product

$$P_{m+1} = f\left(T_1^{m+1}\right) f\left(T_1^m\right) \tag{4.57}$$

is tested. If, as illustrated in the figure, P_{m+1} is negative, then $T_1^m < T_1 < T_1^{m+1}$. When this occurs the search can be terminated and the value T_1^m taken as an acceptable estimate of T_1. Alternatively, the search may be continued between T_1^m and T_1^{m+1} using a smaller value of ΔT_1, say $\Delta T_1/10$.

Now consider the more difficult case where $N > 1$; that is, where more than one surface element has a specified net heat flux. In this case Eq. (4.54) involves a system of N coupled implicit algebraic equations. Standard techniques are available for solving such systems, but perhaps the most robust among these use **genetic algorithms** [10–12]. As the name implies, genetic algorithms are based on an analogy with the natural evolution of species. In addition to being robust and unconditionally stable, they are relatively easy – some would even say fun – to use. Genetic algorithms are a natural choice for engineers and scientists already exploiting the MCRT method.

Briefly, an initial population of potential "parents" is established consisting of individual members whose "genes" are represented by a set of unknown but "guessed" temperatures for the N surfaces for which the heat fluxes are known. The genetic content of each member of the original generation, illustrated in Figure 4.18, is established randomly. Liberal

$T_1^{(1)}$	$T_2^{(1)}$		$T_{i-1}^{(1)}$	$T_i^{(1)}$	$T_{i-1}^{(1)}$		$T_N^{(1)}$	Genes of Parent 1
$T_1^{(2)}$	$T_2^{(2)}$		$T_{i-1}^{(2)}$	$T_i^{(2)}$	$T_{i-1}^{(2)}$		$T_N^{(2)}$	Genes of Parent 2
$T_1^{(3)}$	$T_2^{(3)}$		$T_{i-1}^{(3)}$	$T_i^{(3)}$	$T_{i-1}^{(3)}$		$T_N^{(3)}$	Genes of Parent 3
				...				
$T_1^{(p)}$	$T_2^{(p)}$		$T_{i-1}^{(p)}$	$T_i^{(p)}$	$T_{i-1}^{(p)}$		$T_N^{(p)}$	Genes of Parent p
				...				
$T_1^{(P)}$	$T_2^{(P)}$		$T_{i-1}^{(P)}$	$T_i^{(P)}$	$T_{i-1}^{(P)}$		$T_N^{(P)}$	Genes of Parent P

Figure 4.18 The candidate solution values T making up the genes of the original generation of the P candidate "parents."

upper and lower bounds on temperature are established, usually based on experience, and then values defining the genes of each individual p in the original population are computed as

$$T_i^p = T_{\min} + R_T(T_{\max} - T_{\min}), \tag{4.58}$$

where as usual R_T is the next available random number, and $1 \leq i \leq n$ ranges over the N unknown temperatures.

The next step is to test the genetic content of each member of the original population to see which potential "parents" have the "best" genes. This may be accomplished by introducing each set of candidate surface temperatures into Eq. (4.53) and then computing the corresponding value of a penalty function. A convenient and often used penalty function is the root-mean-square (rms) difference between the left-hand side (LHS) and the right-hand side (RHS) of the N equations,

$$rms_p = \sqrt{\frac{1}{N}\sum_{i=1}^{N}\left(LHS_i^p - RHS_i\right)^2}, \quad 1 \leq p \leq P. \tag{4.59}$$

All members p of the original population are now ranked in ascending order of their rms value; that is, with the parents producing the smallest value of rms_p at the top of the list.

We now arrange "marriages" between randomly paired couples from the upper half of the list (members of the lower half are, alas, not permitted to mate). A marriage consists of using the genetic information of the parents to create "offspring" whose genes are improved in the sense that they will, when tested, produce an even smaller value of rms_p. The simplest approach is to average the two values of each gene provided by the two parents. However, convergence is obtained faster if the parent with the "better" genes is weighted more heavily in the averaging process. Also, experience has shown that judicious use of random "mutations" can lead to more rapid improvement of the gene pool while avoiding "local" minima in the penalty function. In any case, it is best if the marriage leads to two non-identical offspring. This assures that the size of the gene pool will not diminish from one generation to the next, while at the same time maintaining its genetic diversity.

Each successive generation of offspring is treated in the same manner as the original population, with an attendant increase in the quality of the gene pool. This procedure is continued until the relative rms difference between the LHS and the RHS for the best result,

$$rms_p = \frac{\sqrt{\frac{1}{n}\sum_{i=1}^{n}(LHS_i^p - RHS_i)^2}}{\frac{1}{n}\sum_{i=1}^{n}RHS_i}, \quad \text{(best results)} \tag{4.60}$$

is less than some predetermined value, say 0.01. Once the unknown temperatures have been thus obtained, Eq. (4.55) can be used to compute the unknown net heat fluxes on surfaces $N+1, N+2, \ldots, n$.

The advantage of genetic algorithms in this application is that, in contrast to other methods, they do not require numerical estimation of sensitivities, or partial derivatives, of Eq. (4.53) with respect to each of the candidate values. Finally, inverse heat radiation problems might also be solvable using a variation of the approach used to solve inverse heat conduction problems [13], except that now polynomial expressions are used to represent the unknown surface heat flux and temperature distributions rather than their variations with time at fixed points. This idea is currently being pursued by the author and his colleagues.

Problems

4.1 Demonstrate the validity of Eq. (4.6).

4.2 The integrating sphere considered in this chapter offers a rare opportunity to obtain an exact analytical solution using the net-exchange method. Consider the diffuse, gray spherical enclosure illustrated in Figure 4.19. The position vector r with subscripts 1 and 2 in the figure indicates position on the spherical surface; that is,

$$r = xi + yj + zk, \tag{4.61}$$

where $x = R \sin \vartheta \cos \varphi$, $y = R \sin \vartheta \sin \varphi$, and $z = R \cos \vartheta$, with appropriate subscripts. Note that the spherical enclosure has *two* interior surfaces, one from which radiation is leaving and one to which it is arriving.

Let $T(r)$ be the specified wall temperature distribution. Then the integral form for the **radiosity** at point r_1 on the "leaving" surface is

$$\begin{aligned} B(r_1) &= e\sigma T^4(r_1) + (1-e)\int_{A_2} B(r_2)\, dF_{1\to 2} \\ &= e\sigma T^4(r_1) + (1-e)\, H(r_1), \end{aligned} \tag{4.62}$$

where

$$dF_{1\to 2} = \left(\frac{\cos \vartheta_1 \cos \vartheta_2}{\pi L^2}\right) dA(r_2). \tag{4.63}$$

The factor $\left(\frac{\cos \vartheta_1 \cos \vartheta_2}{\pi L^2}\right)$ in Eq. (4.63) is referred to as the **kernel** of the **integral equation**.

(a) Derive Eq. (4.63) using the definition of the differential configuration factor and the definitions of intensity and differential solid angle.

(b) Demonstrate that in the case of the spherical enclosure shown in Figure 4.19,

$$dF_{1\to 2} = \frac{dA(r_2)}{A_2}, \tag{4.64}$$

where $A_2 = 4\pi R^2$ is the area of the "arriving" surface. That is, demonstrate that the integral equation in this case has a constant kernel, $1/A_2$.

(c) Note that with the introduction of Eq. (4.64), Eq. (4.62) becomes

$$B(r_1) = \varepsilon\sigma T^4(r_1) + (1-\varepsilon)\frac{1}{A_2}\int_{A_2} B(r_2)\, dA_2. \tag{4.65}$$

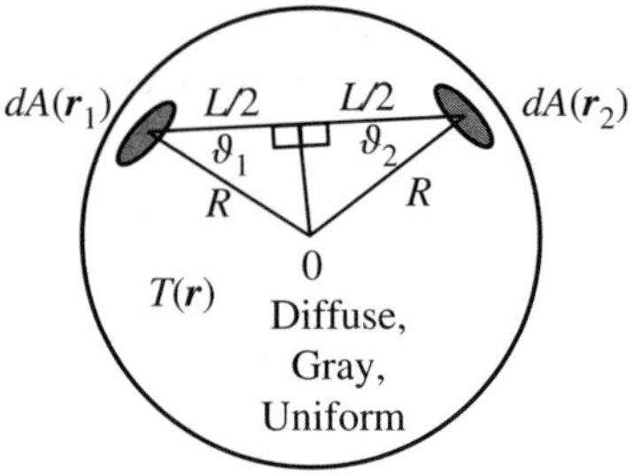

Figure 4.19 A diffuse, gray spherical enclosure (Problem 4.2).

We conclude that the factor multiplying $(1-\varepsilon)$ in Eq. (4.65) must be a constant regardless of the radiosity distribution on A_2 (Explain), and by comparing Eq. (4.65) with Eq. (4.62). We further conclude that this constant must be the **irradiance** $H(r_1)$. In other words, the irradiance on the interior surface of a diffuse, gray spherical enclosure is uniform regardless of the temperature distribution. Furthermore, if the temperature distribution is itself uniform, the radiosity distribution is also uniform. This explains why such enclosures are referred to as "integrating" spheres and why they are valuable for the application explored in this chapter.

(d) Demonstrate that the irradiance on the interior surface of an integrating sphere is

$$H = \frac{1}{A}\int_A \sigma T^4(r)\ dA \tag{4.66}$$

regardless of the value of the absorption coefficient.

4.3 Show that the relative probability that a ray will be emitted at azimuth angle ϑ_e is given by Eq. (4.36), and that this probability is related to a random number R_{ϑ_e} according to Eq. (4.37).

4.4 Derive Eq. (3.16).

4.5 Show that in the case of a diffuse emitter Eq. (4.37) leads to the first expression in Eq. (1.32).

4.6 Derive Eq. (4.39). Hint: Refer back to Section 2.10.

4.7 Assume that a three-column file is available corresponding to a stated wavelength in which each line contains three entries: ϑ_i,

ϑ_r, and $\rho(\lambda, \vartheta_i, \vartheta_r)$. The first column contains ϑ_i values at 5° intervals between 5 and 85° and, for each value of ϑ_i, the second column contains ϑ_r values at 5° intervals between 5 and 85°, and the third column contains the corresponding values of $\rho(\lambda, \vartheta_i, \vartheta_r)$. Write a Matlab function of the form *function[theta_reflected, rho]* = BIDIRECTIONAL(*theta_incident, R_theta*) that uses the file with a random number R_{ϑ_r} to return values of ϑ_r and $\rho(\lambda, \vartheta_i, \vartheta_r)$ for any value of ϑ_i.

4.8 Derive Eqs. (4.43) and (4.44) from the definition of the bidirectional spectral reflectivity, Eq. (4.2).

4.9 Referring back to Figure 1.16, let Surfaces 1–4 be pure platinum and the remaining surfaces be black glass. Create a Matlab code for estimating the band-averaged radiation distribution factor matrix for the enclosure based on the bidirectional reflectivity models developed in Sections 2.12 and 2.13 for the two materials making up the enclosure. Assume that the optical properties n and k are independent of wavelength over the wavelength interval of interest.

(a) Study the convergence of the distribution factors obtained using the program. How many rays must be traced to achieve convergence?

(b) Compare the values of the distribution factors obtained using bidirectional surface optical models with those obtained in Problem 3.20. Interpret any differences obtained.

4.10 Let Surfaces 1–4 be maintained at a temperature of 800 K while the remaining surfaces are maintained at 300 K. Use the radiation distribution factor matrix estimated in Problem 4.9a to compute the surface net heat fluxes. Plot the heat flux distribution on the interior of the enclosure.

4.11 Let the enclosure in Figure 1.16 be a $1-\mathrm{m}^3$ cube. Furthermore, let Surface 1 have a net heat flux of $4.0\,\mathrm{kW\,m^{-2}}$ while the remaining surfaces are maintained at 300 K. What is the temperature of Surface 1?

4.12 For the cubic enclosure in Problem 4.10, let Surfaces 1–4 have a net heat flux of $4.0\,\mathrm{kW\,m^{-2}}$ while the remaining surfaces are maintained at 300 K. Compute the temperatures of Surfaces 1–4 and the net heat fluxes on the remaining surfaces.

References

1. Mahan, J.R. (2002). *Radiation Heat Transfer: A Statistical Approach*. New York: Wiley.
2. Chembar. Lord Aeroglaze Z302 Absorptive Polyurethane – Black. Chembar Inc. https://www.chembar.com/product/lord-aeroglaze-z302-absorptive-polyurethane-black (accessed 1 August 2018).
3. Prokhorov, A. and Prokhorova, N. (2012). Application of the three-component bidirectional reflectance distribution function model to Monte Carlo calculation of spectral effective emissivities of nonisothermal black-body cavities. *Applied Optics* 51 (33): 8003–8012.
4. Mahan, J.R., Vinh, N.Q., and Priestley, K.J., (2018). An application of the Monte Carlo ray-trace method with bidirectional reflection, Paper TFEC-2018-22038, 3rd Thermal and Fluids Engineering Conference, Fort Lauderdale, FL, 4–7 March. (Figures 4.4–4.11 and Table 4.1 reprinted with permission from this paper.)
5. Ulbricht, R. (1900). Die Bestimmung der mittleren räumlichen Lichtintensität durch nur eine Messung. *Elektrotech Zeit* 21: 595–597.
6. Hanssen, L. (2001). Integrating sphere system and method for absolute measurement of transmittance, reflectance and absorptance of specular samples. *Applied Optics* 40 (19): 3196–3204.
7. Hanssen, L.M. and Snail, K.A. (2002). Integrating spheres for mid- and near-infrared reflection spectroscopy. In: *Handbook of Vibrational Spectroscopy*, vol. 2 (ed. J.M. Chalmers and P.R. Griffiths), 1175–1192. New York: Wiley.
8. Mahan, J.R., Walker, J.A., and Stancil, M.M. (2015). Bidirectional reflection effects in practical integrating spheres. *Applied Optics* 54 (30): 8951–8956.
9. Walker, Jr., J.A., (1998). Models and validation measurements of directional reflectance factor for diffuse reflecting materials, PhD Dissertation, University of Arizona.
10. Chambers, L. (1995). *Practical Handbook of Genetic Algorithms*. Boca Raton, FL: CRC Press.
11. Goldberg, D. (1989). *Genetic Algorithms in Search, Optimization, and Machine Learning*. New York: Addison-Wesley.
12. Davis, L. (1991). *Handbook of Genetic Algorithms*. New York: Van Nostrand Reinhold.
13. Samadi, F., Woodbury, K.A., and Beck, J.V. (2018). Evaluation of generalized polynomial function specification methods. *International Journal of Heat and Mass Transfer* 122: 1116–1127.

5

The MCRT Method for Participating Media

So far we have only considered energy interactions between rays and surfaces. Two possible outcomes – absorption and reflection – have been identified, and corresponding models have been formulated and used to describe radiation heat transfer among surfaces. When a ray incident to a surface is neither absorbed nor reflected, it is said to be transmitted. Thermal radiation passing through lenses, filters, and the earth's atmosphere are familiar examples of transmission. This chapter deals with emission, absorption, refraction, and scattering of thermal radiation within a medium through which it is propagating. Such media are said to be "participating."

5.1 Radiation in a Participating Medium

The two previous chapters have dealt with radiation among surfaces, where a surface is defined as the interface between two regions of space having different optical properties. We now turn our attention to the disposition of radiation within volume elements of a **participating medium**; that is, a medium in which radiation is emitted, absorbed,

The Monte Carlo Ray-Trace Method in Radiation Heat Transfer and Applied Optics, First Edition. J. Robert Mahan.

This Work is a co-publication between John Wiley & Sons Ltd and ASME Press.
Companion website: www.wiley.com/go/mahan/monte-carlo

refracted, and scattered. We have already encountered the simplest example of a participating medium, the lens (see Problem 1.14), in which individual rays are redirected by refraction. In addition to refraction, radiation can also be absorbed, emitted, and scattered within a participating medium. In this chapter we learn how to estimate and use radiation distribution factors between volume elements and between volume and surface elements to model radiant exchange in participating media.

Figure 5.1 shows a structured computational fluid dynamics (CFD) mesh adapted by NASA engineers [1] for use in an MCRT model to predict the infrared signature of the Boeing 747 used to ferry the space shuttle between Edwards Air Force Base in California and the Kennedy Space Center in Florida. Figure 5.2 compares the measured mid-wavelength (2–5 μm) infrared image of this aircraft in flight with the predicted image based on the CFD mesh of Figure 5.1. Differences between the predicted

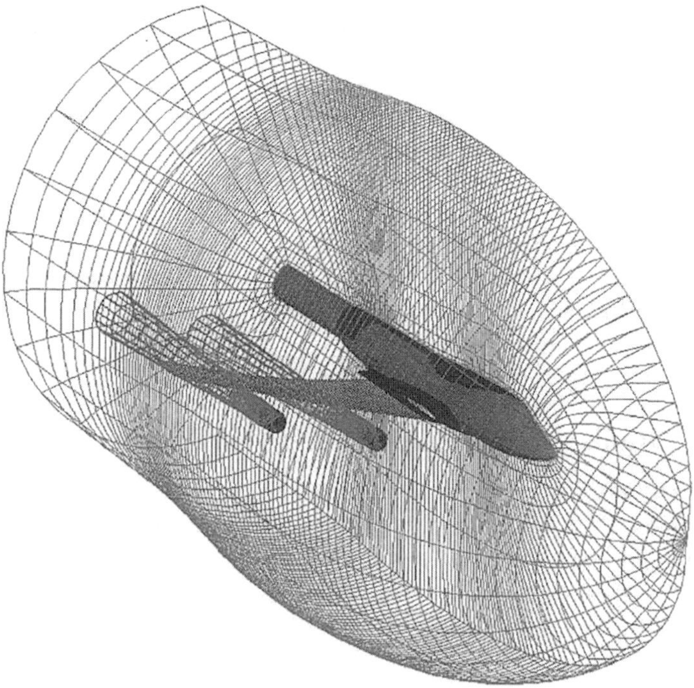

Figure 5.1 Overlay of several structured computational fluid dynamics (CFD) meshes of a Boeing 747. Source: Ref. [1, reprinted with permission]. (*See color plate section for the color representation of this figure*)

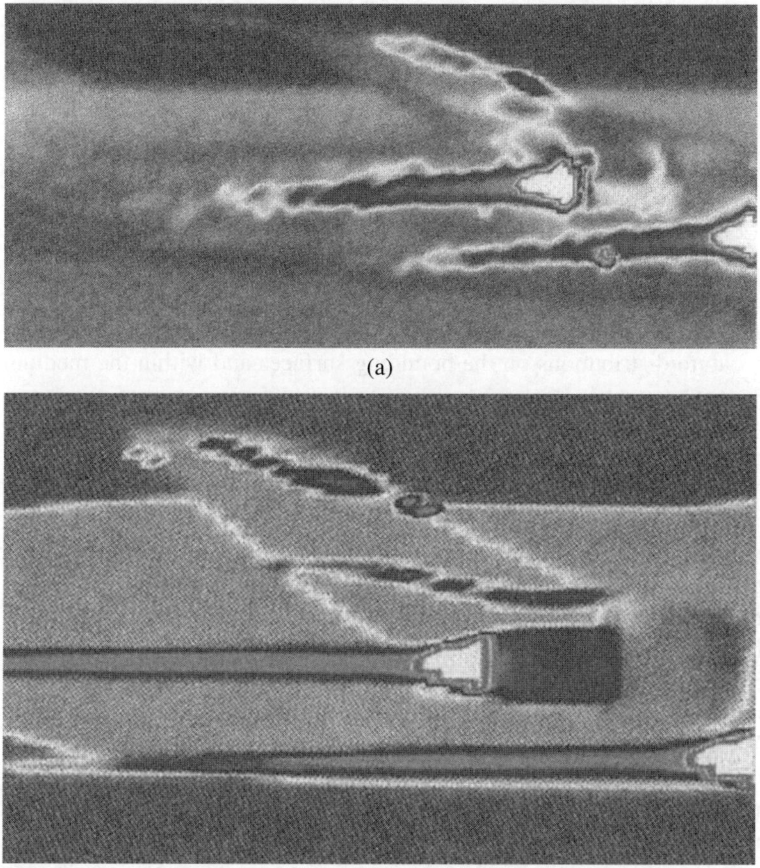

(a)

(b)

Figure 5.2 Comparison of (a) measured and (b) predicted mid-wavelength range infrared images of a Boeing 747 in flight. Source: Ref. [2, reprinted with permission]. (*See color plate section for the color representation of this figure*)

and observed images can be attributed to several factors, but mainly to the limited spatial resolution of the CFD mesh, which contains "only" about 11 million volume elements, or "cells." One obvious departure of the predicted image from the measured image is the length of the engine exhaust plumes, which are too narrow and extend too far downstream in the predicted image. This departure may be attributed to the CFD solution, which provides the temperature and chemical species distributions required by the MCRT model.

The educational goals of the current chapter are to learn how to:

1. Create thermographically true images like the one shown in Figure 5.2b when the bounding surface temperature distributions and distributions of both temperature and chemical species within the medium are known.
2. Obtain the temperature distribution on the bounding surfaces and within a participating medium when the surface heat flux distribution and heat release distribution in the participating medium are known.
3. Obtain the heat flux distribution on the bounding surfaces and the heat release distribution within a participating medium when the temperature distributions on the bounding surfaces and within the medium are known.
4. Obtain solutions for any combination of the situations described in goals 2 and 3.

The required optical properties depend on the concentrations of the various chemical species making up the participating medium. This is especially true when the participating medium is a gas, as in the case of atmospheric radiation and radiation in jet engine combustion chambers and exhaust plumes. Determination of the local optical properties in such cases falls within the realms of fluid dynamics, thermodynamics, and convection heat transfer, and so is beyond the scope of this book. We assume as a starting point that the required optical properties have already been either measured or calculated using an appropriate CFD model. Implicit in this assumption is that radiation heat transfer does not influence the values of the optical properties; that is, that the radiation problem and the fluid dynamics problem are uncoupled. We hasten to add, however, that the method developed in this chapter would be a necessary component of any coupled solution.

5.2 Example: The Absorption Filter

Before undertaking the more general case of gaseous radiation, it is instructive to model the optical performance of a fused silica **absorption filter**. Fused silica filters are commonly used in instruments designed to monitor the earth's radiant energy budget from space, because they permit passage of incident solar radiation while blocking incident

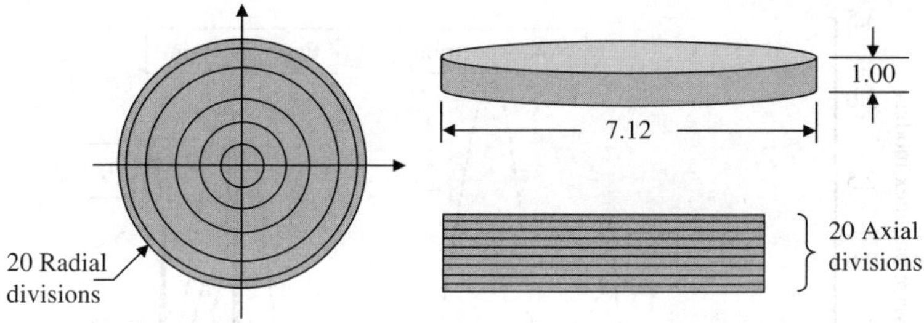

Figure 5.3 A fused silica (silicon dioxide) absorption filter.

earth-emitted radiation. For the purposes of this example, the filter is divided into 400 approximately equal-volume cells by slicing it into 20 equal-thickness axial divisions and 20 equal-area radial divisions, as illustrated in Figure 5.3. The azimuthal divisions shown in Figure 3.5, normally associated with a cylindrical coordinate system, are not needed in this example because the radiation field incident to the filter is assumed to be axisymmetric.

A remarkable compendium of spectral refractive indices, n_λ, and spectral absorption indices, k_λ, of fused silica over a wavelength interval extending from about 0.03 µm to about 1000 µm has been published by Kitamura et al. [3]. In addition to their own data, Kitamura et al. include data from Refs. [4–44]. A running average of these optical properties over a restricted wavelength interval is given in Figure 5.4. The curves are a plot of values for 192 individual wavelengths that are sufficiently closely spaced to permit accurate linear interpolation between adjacent values. Tabulated values are available at the companion website listed on p. xix.

Three phenomena contribute to the optical behavior of an absorption filter:

1. Reflection from the outer surface of the front face
2. Absorption within the filter volume
3. Multiple reflections between the two interior surfaces, as illustrated in Figure 5.5

When a ray intersects the front face of a filter, it is split into two sub-rays, one reflected and one transmitted. In the MCRT environment it is

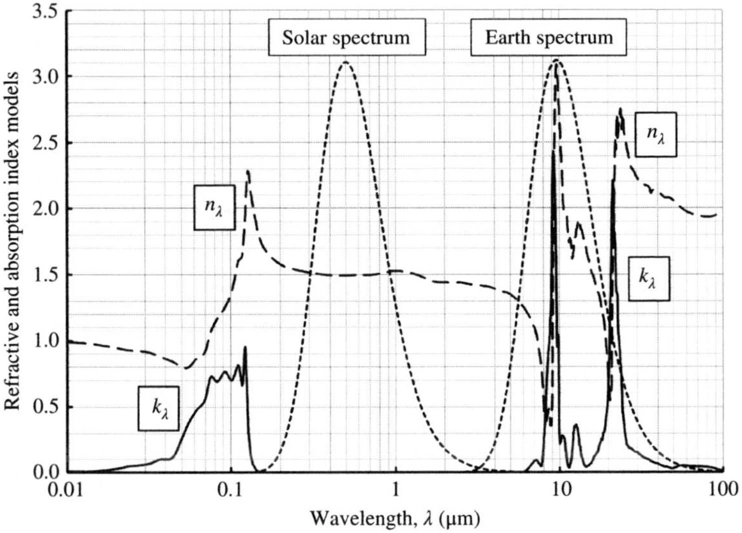

Figure 5.4 Running averages of the optical constant data reported in Ref. 3, with the ideal solar and earth-emitted spectra included. Source: Ref. [45, reprinted with permission].

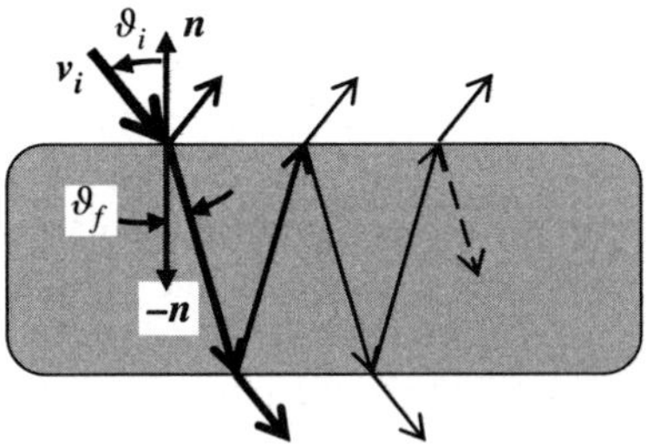

Figure 5.5 Multiple internal reflections within an absorption filter.

usually necessary to trace both rays. Reflection from the front face can be significantly reduced by the use of an anti-reflective (AR) coating, which has the effect of reducing the abruptness with which the refractive index changes from that of a vacuum to that of fused silica. The direction of the reflected ray is computed using Eq. (1.27) and, in the absence of an anti-reflective coating, the Fresnel equation,

$$\rho'_\lambda = \frac{1}{2}\left\{\left[\frac{\cos(\vartheta_i) - \sqrt{n_\lambda^2 - \sin^2(\vartheta_i)}}{\cos(\vartheta_i) + \sqrt{n_\lambda^2 - \sin^2(\vartheta_i)}}\right]^2 + \left[\frac{n_\lambda^2\cos(\vartheta_i) - \sqrt{n_\lambda^2 - \sin^2(\vartheta_i)}}{n_\lambda^2\cos(\vartheta_i) + \sqrt{n_\lambda^2 - \sin^2(\vartheta_i)}}\right]^2\right\}, \tag{5.1}$$

provides an accurate model for the spectral reflectivity of the front face. In Eq. (5.1), $\vartheta_i = \cos^{-1}[\boldsymbol{v}_i \cdot \boldsymbol{n}]$, where $\boldsymbol{v}_i = (L_i, M_i, N_i)$ is the unit vector defining the direction of the ray incident to the front face and $\boldsymbol{n}$ is the front-face unit normal vector, as illustrated in Figure 5.5. Depending on the wavelength, the value of ρ'_λ is typically between 0.02 and 0.04 for fused silica. Application of a good anti-reflective coating can reduce this value to less than 0.01.

Absorption within the filter volume is governed by the "survival equation,"

$$P_\lambda = P_{\lambda,0}e^{-\tau_\lambda}, \tag{5.2}$$

which follows from Beer's law (see Problem 5.1). In Eq. (5.2), $P_{\lambda,0}$ is the power carried by the ray entering the filter and P_λ is the power carried by the ray after traversing the dimensionless **optical distance**,

$$\tau_\lambda \equiv \kappa_\lambda d = \left(\frac{4\pi k_\lambda}{\lambda}\right) d, \tag{5.3}$$

where $\kappa_\lambda = 4\pi k_\lambda/\lambda$ is the **monochromatic absorption coefficient**.

The effect of multiple internal reflections, depicted in Figure 5.5, is negligible outside the filter passband where a practical fused silica filter is said to be **optically thick** ($\tau_\lambda \gg 1$). This means that virtually all of the radiation entering the filter is absorbed during the first pass, in which case the maximum physical length of a ray within the filter is

$$d = t/\cos\vartheta_f. \tag{5.4}$$

In Eq. (5.4), t is the thickness of the filter and $\vartheta_f = \cos^{-1}[-\boldsymbol{n} \cdot \boldsymbol{v}_f]$ is the angle between the ray and the inward normal to the filter front face, as illustrated in Figure 5.5. The unit vector $\boldsymbol{v}_f$ is obtained using the vector form of the Snell–Descartes law, Eq. (1.43).

Internal reflections occur within the passband, where the filter is said to be **optically thin** ($\tau_\lambda \ll 1$), thus allowing the ray to make many passes n through the filter with the same constant attenuation during each pass. In this case the power carried by a transmitted ray leaving the filter is given by

$$P_{\lambda,t} = \sum_{n=1}^{\infty} P_{\lambda,t,n}, \tag{5.5}$$

where it can be demonstrated that its nth-order component is

$$P_{\lambda,t,n} = P_{\lambda,0}(1 - \rho_{ext,\lambda})(1 - \rho_{int,\lambda})(1 - a_\lambda)^{2n-1}\rho_{int,\lambda}^{2(n-1)}. \tag{5.6}$$

In Eq. (5.6), $\rho_{ext,\,\lambda}$ is the reflectivity of the filter front face given by Eq. (5.1), a_λ is the constant effective absorptivity for each pass through the filter, and $\rho_{int,\,\lambda}$ is the reflectivity of the interior interfaces of the filter. We can estimate the importance of internal reflections by computing the relative values of successive orders of internal reflection. Suppose, for example, that $\rho_{ext,\,\lambda} = 0.02$, $a_\lambda = 0.005$, and $\rho_{int,\,\lambda} = 0.02$, which are representative values within the passband of a typical fused silica filter. When we introduce these values into Eq. (5.6), we find that the resulting series converges rapidly with n, as shown in Table 5.1. For a first-order internal reflection ($n = 1$) we find that $P_{\lambda,\,t,\,1}/P_{\lambda,\,0} = 0.9556$, while for a second-order internal reflection ($n = 2$), $P_{\lambda,\,t,\,2}/P_{\lambda,\,0} = 0.000378$. Based on this result we feel confident that we can neglect internal reflections when modeling fused silica filters. The use of an AR coating would further diminish the role of reflections.

Table 5.1 Convergence of Eq. (5.5) when $\rho_{ext} = 0.02$, $a = 0.005$, and $\rho_{int} = 0.02$.

n	$2n-1$	$2(n-1)$	$P_{t,n}$, Eq. (5.6)	P_t, Eq. (5.5)
1	1	0	0.955 598	0.955 598
2	3	2	0.000 378 426	0.955 976
3	5	4	1.49861E−07	0.955 977
4	7	6	5.93463E−11	0.955 977
5	9	8	2.35017E−14	0.955 977

We now turn our attention to the fused silica filter shown in Figure 5.3. It would not be necessary to subdivide the filter into cells if we were only interested in computing its transmittance in a given wavelength interval. However, if we undertake the further task of computing the spatial distribution of absorbed power within its volume, the filter must be subdivided. The distribution of absorbed power in the filter can then be used in a heat conduction model to predict the temperature distribution, which, in turn, can be used to predict the fraction of absorbed power that is re-emitted. Re-emission from the back face is significant because it is equivalent to a long-wavelength "leak" through the filter. An investigation [46] has been carried out to validate the hypothesis that this long-wavelength leak can be significantly reduced by slicing the single 1.0-mm-thick filter into two 0.5-mm-thick filters that are conductively isolated from each other, as illustrated in Figure 5.6. Then the second filter absorbs radiation emitted from the back of the first filter, thereby acting as a radiation shield for the detector.

In the investigation reported in Ref. [46], a population of 59 686 randomly distributed diffuse rays enters the center portion of the filter through the front face. The rays are distributed by wavelength according to the blackbody radiation distribution function representing either earth-emitted radiation or earth-reflected solar radiation. Each ray is traced through the filter for each of the 192 wavelength intervals whose optical constants are tabulated at the companion website listed on p. xix. Therefore, the results obtained really represent an equivalent of 11 459 712 rays passing through the filter.

Figure 5.3 indicates that the filter has been subdivided into 400 cells, each having a volume of approximately 0.1 mm^3. The path of each ray at each wavelength through each cell is determined based on the position and direction of incidence of the ray on the filter front face. Snell's law and the wavelength dependence of the index of refraction assure that

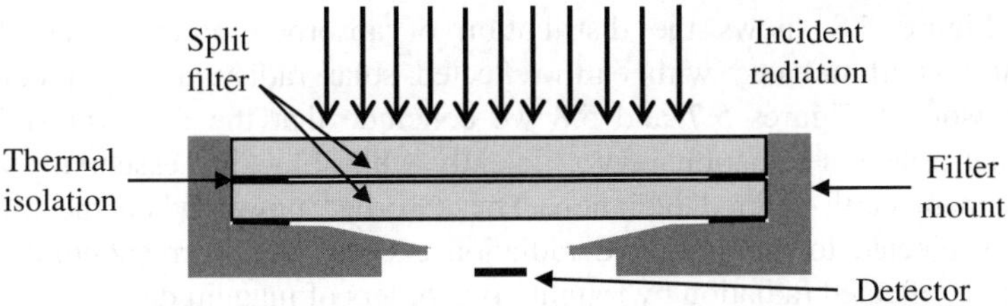

Figure 5.6 Split-filter design for reducing long-wavelength "leak."

the path through the filter of any one ray is different at each wavelength. Therefore, the calculation for each of the 59 686 rays at each of the 192 wavelengths through each of the 400 cells comprising the filter constitutes the solution of 4.58×10^9 absorption problems, where each problem consists of computing the amount of power left behind in a given cell by absorption due to a given ray and a given wavelength interval; that is,

$$P_{\lambda,abs} = P_{\lambda,inc}(e^{-4\pi k_\lambda L_1/\lambda} - e^{-4\pi k_\lambda L_2/\lambda}). \tag{5.7}$$

In Eq. (5.7), $P_{\lambda,\,inc}$ is the power (W) entering the cell after the ray has traveled a distance L_1 in the filter, and L_2 is the distance the ray has traveled in the filter as it exits the cell; that is, $L_2 - L_1$ is the distance the ray has traveled through the filter. The absorption index k_λ is the mean value on the wavelength interval under consideration, as is the refractive index n_λ used to steer the ray through the filter. Reflections from the front face of the filter are ignored; that is, the use of an effective anti-reflective coating is assumed.

The spectral distributions for the simulated earth-emitted radiation and earth-reflected solar radiation are represented in Figure 5.4. In addition to the presence of a good anti-reflective coating, we assume that the radiation has already passed through a collimator and a reflecting telescope stage arranged as illustrated in Figure 3.2. The results of the simulation are given in Figures 5.7 and 5.8. The total power incident to the front face of the filter is approximately the same in the two cases.

Figure 5.7 is a plot of the distribution of absorbed earth-emitted radiation within the filter. Shown are the radial distributions of power absorbed in the two layers of cells nearest the front face of the filter. We see that more than 99% of the incident power is absorbed, mostly in the layer nearest the front face. This produces a steep front-to-back temperature gradient that conducts the absorbed power deeper into the filter, leading to a potentially significant temperature rise on the back face. It is this possibility that encourages us to consider splitting the filter into two halves.

Figure 5.8 shows the distribution of absorbed power when the filter is illuminated with earth-reflected solar radiation. Upon comparison of Figures 5.7 and 5.8 we conclude that the filter transmits solar-reflected radiation very efficiently while blocking essentially all incident earth-emitted radiation. The absorbed power when the filter is subjected to earth-emitted radiation exceeds that corresponding to solar-reflected radiation by roughly two orders of magnitude.

We conclude that the value of radiation transmitted through the filter must greatly exceed the value of power re-radiated from the back

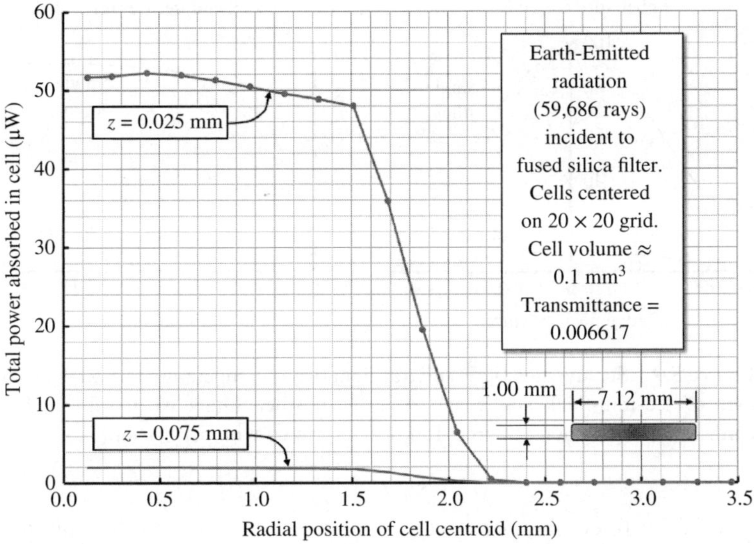

Figure 5.7 Distribution of absorbed earth-emitted radiation in the fused silica filter.

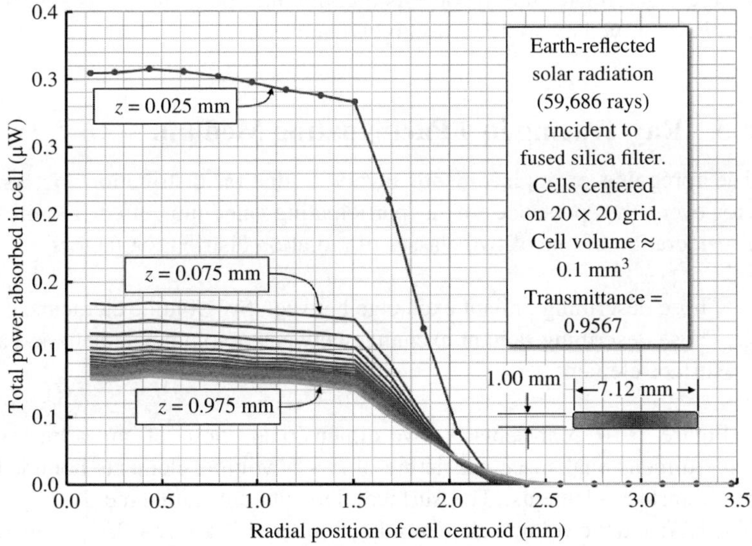

Figure 5.8 Distribution of absorbed solar-reflected radiation in the fused silica filter.

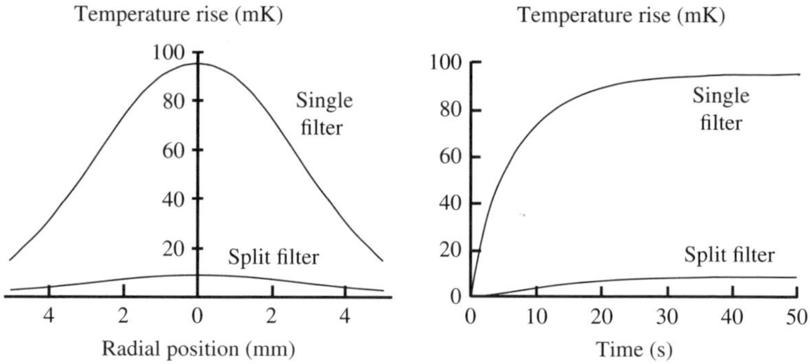

Figure 5.9 Steady-state (left) and transient (right) temperature rise on the back face of 1.0-mm-thick single and split fused silica filters when subjected to earth-emitted radiation. Source: Ref. [46, reprinted with permission].

face of the filter when observing solar-reflected radiation, but that this might not be true when observing earth-emitted radiation. Therefore, we employ a transient conduction analysis, described in Ref. [46], to compute the transient temperature rise on the back face of the filter. Study of Figure 5.9 validates the hypothesis that the filter performance is significantly improved by splitting it into two halves.

5.3 Ray Tracing in a Participating Medium

The foregoing example did not involve radiation distribution factors. However, the presence of a participating medium often requires consideration of one of two classes of radiation distribution factors:

1. Those describing radiant exchange between two volume elements.
2. Those describing radiant exchange between a volume element and a surface element.

Figure 5.10 represents a two-dimensional view of an arbitrary three-dimensional space consisting of $n - N$ volume elements bounded by N surface elements. The surface elements are numbered 1, 2, …, $N - 1, N$; and the volume elements are numbered $N + 1, N + 2, \ldots, n - 1$, n. The indices i and j are used to designate any pair of elements, surface, or volume, with $1 \leq i \leq n$ and $1 \leq j \leq n$, including the possibility that $i = j$.

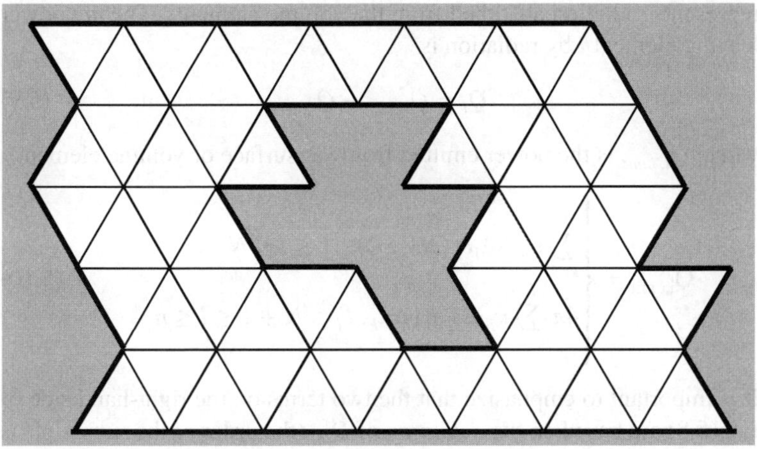

Figure 5.10 Two-dimensional representation of a notional three-dimensional enclosure containing a participating medium. White space is participating medium, gray space is solid, heavy lines are surfaces, and light lines are boundaries between volume elements.

Either the net heat flux ($\mathrm{W\,m^{-2}}$) or the temperature (K) may be specified on any surface, and either the net volumetric cooling rate ($\mathrm{W\,m^{-3}}$) or the temperature (K) may be specified for any volume element. All optical properties are assumed constant and known, although spatial and spectral variations of their values are allowed. The index k is used to designate the wavelength interval $\Delta\lambda_k$ corresponding to a specified value of an optical property. Care should be taken to avoid confusing the index i with the blackbody intensity i_b, and the index k with the absorption *coefficient* κ_λ and the absorption *index* k_λ.

From the definition of the radiation distribution factor D_{ijk}, it follows that the power (W) absorbed by (surface or volume) element i is

$$Q_{i,abs} = \sum_{k=1}^{K} \sum_{j=1}^{N} \varepsilon_{jk}\, \Delta A_j\, e_b(\Delta\lambda_k, T_j)\, D_{jik} + 4\pi \sum_{k=1}^{K} \sum_{j=N+1}^{n} \kappa_{jk}\, \Delta V_j\, i_b(\Delta\lambda_k, T_j)\, D_{jik}. \tag{5.8}$$

The first term on the right-hand side of Eq. (5.8) represents radiation absorbed from the surface elements of the enclosure, and the second term represents radiation absorbed from the volume elements. The *net* power leaving element i by radiation is

$$Q_i = Q_{i,emit} - Q_{i,abs}, \tag{5.9}$$

where $Q_{i,\,emit}$ is the power emitted from the surface or volume element,

$$Q_{i,emit} = \begin{cases} \sum_{k=1}^{K} \varepsilon_{ik} \Delta A_i e_b(\Delta\lambda_k, T_i), & 1 \le i \le N \\ 4\pi \sum_{k=1}^{K} \kappa_{ik} \Delta V_i i_b(\Delta\lambda_k, T_i), & N+1 \le i \le n \end{cases} \tag{5.10}$$

It is important to emphasize that the two terms on the right-hand side of Eq. (5.8) are based on the *definition* of D_{jik} (the order of the subscripts is important).

As in the case of surface-to-surface formulations studied in the previous chapters, we once again use the MCRT method to estimate the surface-to-volume (or volume-to-surface) and volume-to-volume radiation distribution factors. Figure 5.11 shows the first six steps used to compute the distribution factors from a single element. The process begins after the surface or volume element has been designated, which would be Step 0 (not shown). The order of selection in this step follows the order in which the elements have been numbered, or indexed.

***Step 1*: Determine the emission site.**

The emission site can lie either on a surface element or within a volume element. In both cases emission is into the participating medium. As a concession to pedagogical clarity we consider rectangular-solid volume elements, as illustrated in Figure 5.12. The actual form of the volume

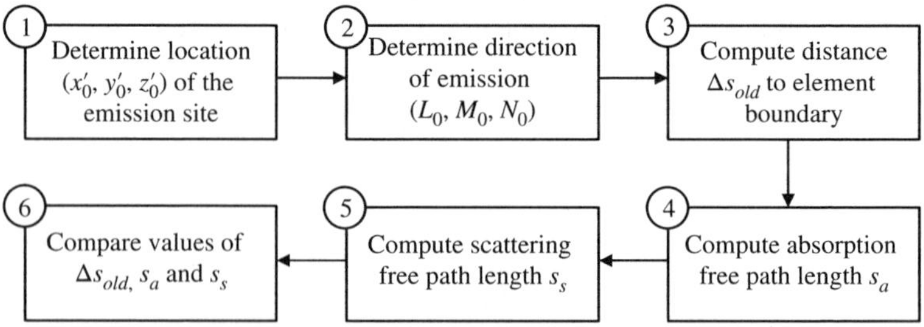

Figure 5.11 The initial steps for computing the distribution factors from one surface or volume element of an enclosure filled with a participating medium.

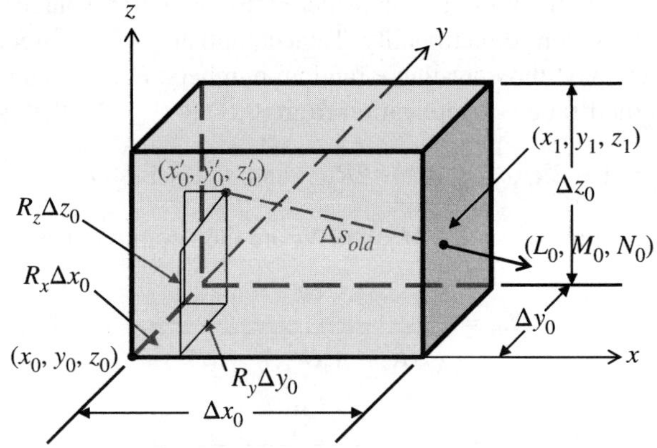

Figure 5.12 A ray emitted within a rectangular-solid volume element filled with a participating medium.

elements will of course be dictated by the enclosure geometry. Tetrahedral volume elements are often used in unstructured meshes because they conform well to irregular geometries.

Let x_0, y_0, and z_0 be the coordinates of the corner of a rectangular-solid volume element whose dimensions are Δx_0, Δy_0, and Δz_0. In Matlab a *function*

$$\text{function } [x_0, y_0, z_0] = \text{CORNER}(i)$$

is usually invoked that returns the corner coordinates of volume element i. The function could be based on a look-up table, but it is usually possible to express x_0, y_0, and z_0 as functions of the index i, as described in Chapter 1. When the emission site lies on a surface element, it is treated as a special case of an interior emission site lying on one of the bounding surfaces of a volume element. We draw the next three available random numbers; R_x, R_y, and R_z, uniformly distributed between zero and unity. Then the coordinates of the emission site are

$$x_0' = x_0 + R_x\Delta x_0, \quad y_0' = y_0 + R_y\Delta y_0, \quad \text{and} \quad z_0' = z_0 + R_z\Delta z_0. \tag{5.11}$$

In the case of a surface element, only two random numbers are required, as established in Chapter 3.

***Step 2*: Determine the direction of emission.**

Later on when we consider scattering we will see that scattering sites act as directional point sources. However, emission sites within a participating medium are considered to be point sources having completely random directionality. To accomplish random directionality we draw the next three available random numbers, R_L, R_M, and R_N, and then shift their range of application from (0, 1) to (−1, 1); that is,

$$R'_L = 1 - 2R_L,\ \ R'_M = 1 - 2R_M,\ \ \text{and}\ \ R'_N = 1 - 2R_N. \tag{5.12}$$

The direction cosines L_0, M_0, and N_0 are then computed as

$$L_0 = \frac{R'_L}{\sqrt{(R'_L)^2 + (R'_M)^2 + (R'_N)^2}}, \tag{5.13a}$$

$$M_0 = \frac{R'_M}{\sqrt{(R'_L)^2 + (R'_M)^2 + (R'_N)^2}}, \tag{5.13b}$$

and

$$N_0 = \frac{R'_N}{\sqrt{(R'_L)^2 + (R'_M)^2 + (R'_N)^2}}. \tag{5.13c}$$

If the emission site lies on a surface element and emission is diffuse, we compute the direction cosines as described in Section 1.4 using Eq. (1.34), or as described in Step 8 of Section 3.3. Note that in the former approach this would require that the surface unit normal vector $\boldsymbol{n}$ and the two surface unit tangent vectors $\boldsymbol{t}_1$ and $\boldsymbol{t}_2$ be known or computed. Finally, if the emission site lies on a directionally emitting surface, we must proceed as described in Step 2 of Section 4.5.

***Step 3*: Compute the distance Δs_{old} from the emission site to the bounding surface in the direction of emission.**

The rectangular-solid volume element illustrated in Figure 5.12 has six faces: one at $x = x_0$, one at $x = x_0 + \Delta x_0$, one at $y = y_0$, one at $y = y_0 + \Delta y_0$, one at $z = z_0$, and one at $z = z_0 + \Delta z_0$. Eq. (3.19) can be rewritten in terms of the nomenclature of the current chapter, to obtain

$$t_1 = \frac{n_x(x_0 - x'_0) + n_y(y_0 - y'_0) + n_z(z_0 - z'_0)}{n_x L_0 + n_y M_0 + n_z N_0}. \tag{5.14}$$

Replacing x_0, y_0, and z_0 respectively with $x_0 + \Delta x_0$, $y_0 + \Delta y_0$ and $z = z_0 + \Delta z_0$ when appropriate, results in six candidate values of t_1. We test all six values, rejecting the three negative values as corresponding to back candidates and retaining the smallest of the three positive values, which we recognize as being the distance Δs_{old} from the emission site to the bounding surface in the direction of emission.

Although intersections along edges or in corners are highly unlikely when millions of rays are traced through hundreds of elements, this is bound to happen occasionally, at least to within the accuracy of the processor being used. Therefore, the logic used to implement this step must take these possibilities into consideration. Another unlikely but possible scenario is that an energy bundle will be emitted, to within the calculation accuracy of the processor being used, exactly parallel to one or even to two of the faces. In this case the program logic must be able to deal with the possibility of one or even two zero direction cosines.

Once the smallest positive value of t_1 has been identified, we can also compute the coordinates for the point of intersection on the wall of the volume element,

$$x_1 = x_0' + L_0 t_1, \quad y_1 = y_0' + M_0 t_1 \quad \text{and} \quad z_1 = z_0' + N_0 t_1, \tag{5.15}$$

and, should it become relevant, the index of the surface (or volume) element intersected.

An aside on meshing

Before continuing to Step 4, we should consider an important implication of absorption of an energy bundle within a participating medium. We have already encountered the idea of the optical distance τ_λ in Section 5.2. If d in Eq. (5.3) is replaced by Δs_1 such that $\tau_\lambda = (4\pi k_\lambda/\lambda)\Delta s_1$, and if the resulting value of τ_λ is large, say $\tau_\lambda > 10$, then we conclude that the ray is **self-absorbed**. When the rays emitted within a cell are likely to be self-absorbed, the cell is said to be **optically thick**. If an enclosure containing a participating medium is meshed such that the resulting volume elements are optically thick for the wavelength interval of interest, then only radiation emitted by volume elements near the enclosure walls will penetrate to the walls. This situation leads to gross inefficiencies in the MCRT implementation and unnecessarily compromises the accuracy of the resulting model. This is also true if the meshing scheme yields **optically thin** volume elements; that is, elements for which $\tau_\lambda \ll 1$. Experience has verified the intuitively attractive idea that the meshing scheme should produce volume elements that are neither optically thick nor optically thin in the wavelength interval of interest. That is, if Δ is the characteristic dimension of a typical volume element, then the meshing scheme should produce

$\tau_\lambda = \kappa_\lambda \Delta \approx 1$. For cube-like cells whose sides are approximately the same length, $\Delta = (\Delta_x + \Delta_y + \Delta_z)/3$; while for rectangular-solid elements whose sides have slightly different lengths, $\Delta = (\Delta_x\ \Delta_y\ \Delta_z)^{1/3}$ might be more appropriate. It is best practice to avoid meshing schemes for which Δ_x, Δ_y, and Δ_z are significantly different from each other, because this leads to situations for which cells are optically thick in one direction and optically thin in another.

It is a challenge to use optimally sized volume elements when the value of the spectral absorption coefficient k_λ varies widely over the spectral range of interest, as it does, for example, near the edges of the pass-band for fused silica in Figure 5.4. One approach is to build up the volume elements corresponding to the more weakly absorbing bands from smaller "building block" volume elements sized for the strongest absorbing bands. Then the number of these building blocks used to form a given larger volume element would vary with wavelength interval. This is similar to the block-structured meshing schemes used extensively in CFD applications, except that now the block structure is determined by wavelength rather than by local turbulence scale. This approach evidently remains to be more fully explored in the literature.

***Step 4*: Compute the absorption mean free path of the energy bundle.**

Consider a beam having a local value of monochromatic intensity $i_{\lambda,0}(\lambda, s)$ propagating at location s in an absorbing medium consisting of a single chemical species n. The local **single-species monochromatic absorption coefficient** $\kappa_{\lambda,n}(\lambda, s)$ is defined by **Beer's law**,

$$di_{\lambda,n,a}(\lambda, s) = -\kappa_{\lambda,n}(\lambda, s) i_{\lambda,0}(\lambda, s) ds. \tag{5.16}$$

The value of $\kappa_{\lambda,n}(\lambda, s)$ is normally obtained experimentally, although absorption in gases is amenable to theoretical treatment based on first principles. When multiple chemical species are present, as is usually the case in gaseous radiation problems, we cannot simply sum the contributions of the individual species, that is,

$$di_{\lambda,a}(\lambda, s) \neq \sum_{n=1}^{N} di_{\lambda,n,a}(\lambda, s), \tag{5.17}$$

where N is the number of species present. This is because the various molecules present interact with each other to an extent that depends on their individual partial pressures and the mixture temperature. That is, the value of $\kappa_{\lambda,n}(\lambda, s)$ for species n depends on the partial pressures of all N of the species present. The situation is further complicated when chemical reactions are present, in which case the local distribution of chemical species is governed by local heat release or absorption through thermodynamics. The treatment offered here is, therefore, limited to absorbing media consisting of a single species or to non-reacting mixtures, such as the jet exhaust plumes depicted in Figure 5.2, for which the multispecies monochromatic absorption coefficient is a known function of temperature and pressure.

It follows from Eq. (5.16) that the **spectral absorptance** $a_{\Delta\lambda}(s_a)$ corresponding to the path length s_a of an energy bundle whose energy is confined to wavelength interval $\Delta\lambda$ is

$$a_{\Delta\lambda}(s_a) = 1 - e^{-\kappa_{\Delta\lambda} s_a}. \tag{5.18}$$

In keeping with the tenets of the MCRT method, we interpret the spectral absorptance as the probability that the energy bundle will be absorbed after traveling a distance s_a. Therefore, we equate $a_{\Delta\lambda}$ to the next available random number R_a and rearrange Eq. (5.18) to obtain the **absorption mean free path** for the energy bundle,

$$s_a = -\frac{1}{\kappa_{\Delta\lambda}} \ln(1 - R_a). \tag{5.19}$$

***Step 5*: Compute the scattering mean free path of the energy bundle.**

The **monochromatic scattering coefficient** $\sigma_\lambda(\lambda, s)$ is defined by a model analogous to Beer's law,

$$di_{\lambda,s}(\lambda, s) = -\sigma_\lambda(\lambda, s) i_{\lambda,0}(\lambda, s) ds_s. \tag{5.20}$$

In Eq. (5.20), $di_{\lambda,s}(\lambda, s)$ is the differential change in monochromatic intensity at location s as it traverses a differential distance ds in a scattering medium where the local value of the monochromatic scattering coefficient is $\sigma_\lambda(\lambda, s)$. In a process similar to the one used to derive Eq. (5.19), Eq. (5.20) can be manipulated to relate the distance this particular energy bundle can be expected to travel before being scattered to the next available random number R_s,

$$s_s = -\frac{1}{\sigma_{\Delta\lambda}} \ln(1 - R_s). \tag{5.21}$$

***Step 6*: Compare the absorption mean path length and the scattering mean path length with each other and with the distance to a bounding surface in the direction of emission.**

We are now confronted with three possible dispositions for the current energy bundle:

1. it exits the current volume element without being absorbed or scattered, or
2. it is absorbed before being either scattered or exiting the current volume element, or
3. it is scattered before being either absorbed or exiting the current volume element.

Note that in the third eventuality, it might still be absorbed in the current volume element after first being scattered. If both s_a and s_s are greater that Δs_{old}, we continue to Step 7, which takes the energy bundle either into a neighboring "new" volume element or to a bounding surface, as appropriate. Alternatively, if s_a is less than both s_s and Δs_{old}, the energy bundle is absorbed in the current volume element. In this case we continue to Step 8, in which the ray count for the current volume element and the total ray count are both incremented. If the total ray count is less than a maximum value, we return to Step 1; otherwise, we terminate the ray trace. Finally, if s_s is less than both s_a and Δs_{old}, the energy bundle is redirected in the current volume element, in which case we continue to Step 9. The logic block diagram begun in Figure 5.11 is continued in Figure 5.13.

***Step 7*: Determine whether the boundary at Δs_{old} is a wall segment or a cell interface.**

Note that an energy bundle can follow one of two possible paths to this step:

1. the direct path from Step 6 to Step 7, as illustrated in Figure 5.13;
2. an indirect path followed after being redirected by scattering in Step 9.

Consider first the case where the energy bundle has reached the boundary of the original ("old") volume element without being absorbed. This

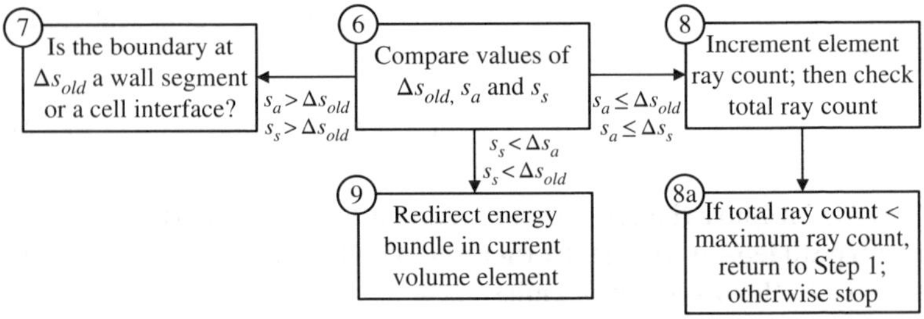

Figure 5.13 Logic followed after comparing the values of Δs_{old}, s_a and s_s.

boundary could be either (i) an interface between two adjacent volume elements or (ii) a wall segment of the enclosure. But how do we determine which? Recall that all of the wall segments and cells have been pre-numbered. We first determine the number, or index i, of the surface or volume element whose boundary lies at Δs_{old}. Then if $1 \leq i \leq N$, the boundary is a surface element, and if $N+1 \leq i \leq n$, the boundary is an interface with another volume element.

The process of determining the index number of a surface or volume element, called "indexing," has already been developed in Section 1.6. If the enclosure geometry can be accurately represented using a combination of rectangular and right-cylindrical solids, the index number can be expressed as an explicit function of the coordinates of a point on a surface or within a volume element. A simple but common example would be a rectangular solid enclosure consisting of rectangular wall segments and rectangular-solid volume elements. Figure 5.14 represents an interior volume element whose corner is located at (x_p, y_q, z_r), where p, q, and r are integer indices whose ranges are $1 \leq p \leq P$, $1 \leq q \leq Q$, and $1 \leq r \leq R$. Then the corner of volume element $N+1$, where N is the number of surface elements, is located at (x_1, y_1, z_1), and the corner of volume element n is located at (x_P, y_Q, z_R). Furthermore, if the x-, y-, and z-ranges of the enclosure are $0 \leq x \leq X$, $0 \leq y \leq Y$, and $0 \leq z \leq Z$, respectively, then $\Delta x = X/P$, $\Delta y = Y/Q$, and $\Delta z = Z/R$. It follows that the pseudo-code expression for the index of the cell containing point (x, y, z) is

$$i = i\_x + \mathrm{i\_y} + i\_z, \tag{5.22}$$

with

$$\text{if } i\_x > 0, i\_x = floor(P^*x/X); \text{else } i\_x = 1, \tag{5.23a}$$

$$\text{if } i\_y > 0, i\_y = floor(Q^*y/Y); \text{else } i\_y = 1, \tag{5.23b}$$

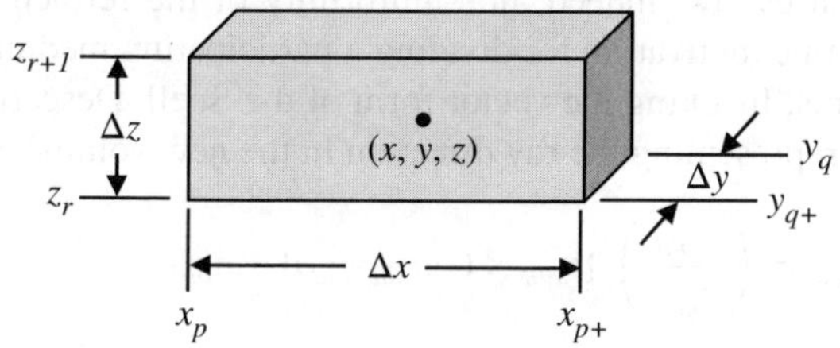

Figure 5.14 A typical volume element, or cell, of a rectangular solid enclosure containing a participating medium.

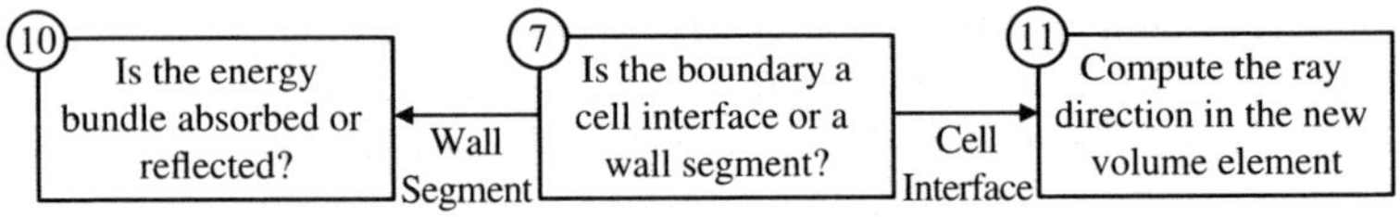

Figure 5.15 Logic following identification of the boundary as either a cell interface or a wall segment.

$$\text{if } i\_z > 0, i\_z = floor(R^* z/Z); \text{else } i\_z = 1. \tag{5.23c}$$

Recall from Chapter 1 that the *floor*(a) operator in Eqs. (5.23) returns the integer value of the argument a by truncation after the decimal point, with the special case that $floor(0) = 0$. An analogous process leads to similar expressions for the indices of surface elements 1 to N. In the case of simultaneous intersection with more than one element, the index number of the first element identified is arbitrarily retained.

The logic following Step 7 is continued in Figure 5.15.

***Step 10*: Determine whether the energy bundle is absorbed or reflected by the wall segment.**

The boundary has been determined to be a surface element because $1 \leq i \leq N$. An $N \times 11$ matrix, Surfaces = [*i*, *alpha*, *n_x*, *n_y*, *n_z*, *t1_x*, *t1_y*, *t1_z*, *t2_x*, *t2_y*, *t2_z*], has been pre-defined containing the absorptivity of surface element i as well as its surface normal and tangent vectors. We draw the next available random number R_α and compare its value to the absorptivity α of surface element i. If $\alpha \geq R_\alpha$ the ray is absorbed (Step 8); otherwise it is reflected (Step 15).

***Step 11*: Compute the ray direction in the new volume element.**

The energy bundle has entered a new volume element from its neighbor. In general, the index of refraction will change from one cell to the next, thereby imposing a change in direction of the ray as required by the Snell–Descartes law. Indeed, non-uniformity of the refractive index is often the prime motivation for dividing a participating medium into volume elements. Invoking the vector form of the Snell–Descartes law, the unit vector representing the ray direction in the new volume element is

$$v_{\mathbf{new}} = \left(\frac{n_{old}}{n_{new}}\right) [n_{old} \times (-n_{old} v_{old})] - n_{old}$$
$$\times \left[1 - \left(\frac{n_{old}}{n_{new}}\right)^2 (n_{old} v_{old})\,(n_{old} v_{old})\right]^{1/2}, \tag{5.24}$$

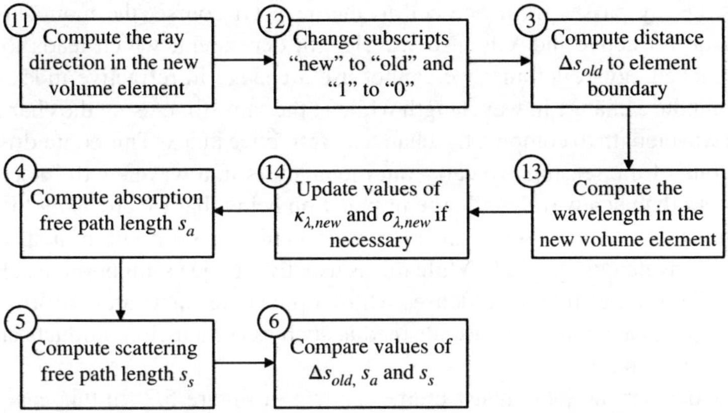

Figure 5.16 Logic flow after computing the ray direction in the new volume element.

where $\boldsymbol{n}_{old}$ is the unit normal vector of the boundary directed into the old volume element, n_{old} and n_{new} are the indices of refraction in the old and new volume elements, and

$$v_{old} = L_0\,\mathbf{i} + M_0\,\mathbf{j} + N_0\,\mathbf{k} \tag{5.25}$$

is the unit vector representing the direction of the ray in the old volume element. The logic flow beyond Step 11 continues in Figure 5.16.

***Step 12*: Change "new" subscripts to "old" and "1" subscripts to "0".**

The ray has entered a new volume element at point (x_1, y_1, z_1) and with direction $v_{new} = (L_1, M_1, N_1)$, so we change the "new" subscript to "old" and the "1" subscript to "0" and return to Step 3.

***Step 13*: Compute the wavelength in the new volume element.**

A ray can be viewed as having a wavelength λ and a corresponding frequency ν, a concept explored more fully in Chapter 6. As a ray whose frequency is ν passes from a cell having a refractive index of n_{old} to one having a refractive index of n_{new}, its wavelength changes from $\lambda_{old} = c_0/n_{old}\nu$ to $\lambda_{new} = c_0/n_{new}\nu$, where c_0 is the speed of light in a vacuum. Thus,

$$\lambda_{new} = (n_{old}/n_{new})\lambda_{old}. \tag{5.26}$$

***Step 14*: Update the values of $\kappa_{\lambda,new}$ and $\sigma_{\lambda,new}$ if necessary.**

The monochromatic absorption and scattering coefficients are generally wavelength dependent (see Figure 5.4) and, therefore, can change as the ray passes from one cell to the next. Of course, the monochromatic refractive index is also wavelength dependent, which leads to a "chicken-egg" dilemma: we cannot use a change in refractive index to compute a change in wavelength while at the same time using the change in wavelength to compute the change in refractive index. This conundrum is one of the reasons we carry out the analysis in a wavelength interval rather than at a wavelength, the implication being that we are using local optical constants whose values are reasonably constant within a specified wavelength interval. While this is usually a good assumption in solid media, such as filters and lenses whose optical constants are uniform, it is often problematic in gaseous media, such as combustion products and exhaust plumes.

Consider the jet exhaust plumes shown in Figure 5.2. In this case it is acceptable to first compute the distributions of pressure, temperature, and chemical species using a CFD code that ignores thermal radiation, and to then use the results obtained to compute the infrared image using an MCRT code. This works because thermal radiation from the plume is too weak to influence the CFD problem. A recent contribution by Ren et al. [47] uses the MCRT method to investigate the influence of radiation on the temperature and species distributions within a gas turbine combustor.

An aside on gaseous radiation

Equation (5.26) can be rearranged to obtain

$$\frac{\lambda_{new} - \lambda_{old}}{\lambda_{old}} = \frac{\Delta\lambda}{\lambda_{old}} = \frac{n_{old}}{n_{new}} - 1. \qquad (5.27)$$

Even though our MCRT model is strictly valid only for the wavelength interval $\Delta\lambda_k$, the wavelength shift occasioned by the change in index of refraction from one volume element to the next may take us outside this interval. Nevertheless, we continue tracing the energy bundle using local spectral absorption coefficients based on the local wavelength. The wavelength interval $\Delta\lambda_k$ occupied by a ray is then taken to be the spectral interval it occupied at its original emission.

For a gas, the index of refraction n is proportional to the mass density ρ (kg m^{-3}). Therefore,

$$\frac{\Delta\lambda}{\lambda_{old}} = \frac{\Delta\rho/\rho_{old}}{1 + \Delta\rho/\rho_{old}} \text{ (gas)}, \qquad (5.28)$$

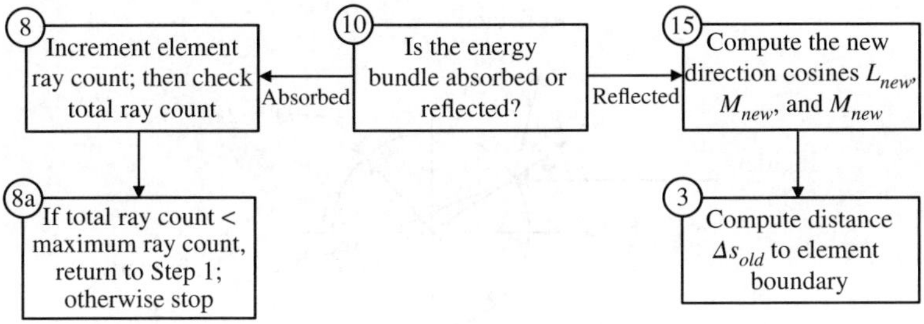

Figure 5.17 Disposition of a ray incident to a surface element.

and for an ideal gas ($p = \rho RT$) at constant pressure, $\Delta\rho/\rho_{old} = -\Delta T/T_{old}$. Therefore,

$$\frac{\Delta\lambda}{\lambda_{old}} = \frac{\Delta T/T}{1 - \Delta T/T_{old}} \text{ (ideal gas).} \tag{5.29}$$

We conclude that a 10% increase in temperature of an ideal gas over the life cycle of an energy bundle would translate into an 11% increase in wavelength.

Whether or not Steps 13 and 14 are executed, we continue on to Steps 4–6.

***Step 15*: Compute the new direction cosines, L_{new}, M_{new}, and N_{new}.**

The logic flow beyond Step 10 is continued in Figure 5.17. Step 15 is identical to Step 6 in Figure 3.1 of Chapter 3 if a diffuse-specular surface reflectivity model is being used. If a directional reflectivity model is being used, we follow the procedure described in Chapter 4.

An aside on scattering

Scattering in a participating medium is analogous to reflection, the main difference being that it occurs in 4π-space rather than in 2π-space. Scattering centers can have any shape but are usually modeled as semi-transparent spheres whose diameters are very small compared to the enclosure dimensions. Examples include particulates produced by combustion (i.e., smoke), dust storms, breaking ocean waves, and volcanism. The two predominant scattering theories are attributed to Lord Rayleigh [48] and Gustav Mie [49]. Detailed developments of these two theories may be found elsewhere [50, 51]. Rayleigh scattering, which describes scattering by molecules, is useful in atmospheric radiation studies, whereas Mie scattering, which describes scattering

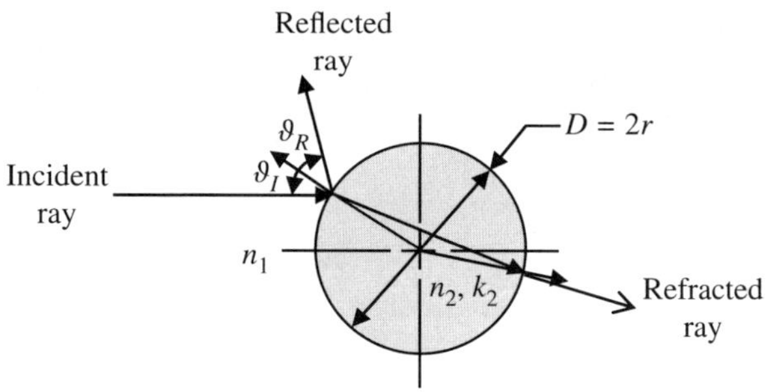

Figure 5.18 Illustration of a MCRT model for scattering by a semi-transparent spherical particle with $n_2 > n_1$ and $2\pi r/\lambda \gg 1$.

by larger particulates such as aerosols, is useful in combustion and atmospheric pollution studies.

In principle, scattering can be directly modeled using the MCRT method if the wavelength is sufficiently short compared to the size of the scattering center, as suggested by Figure 5.18. A large number of collimated rays filling the cross-section of the scattering center are incident from a prescribed direction. Note that absorption can occur as the transmitted rays transit the scattering center. Reflections are considered to be mirror-like ($\vartheta_I = \vartheta_R$), with a reflectivity given by the Fresnel equation, Eq. (5.1). Refraction is governed by the Snell–Descartes law, Eq. (5.24), and internal absorption is treated as described above. This numerical simulation of scattering yields both a monochromatic scattering coefficient σ_λ and the **monochromatic scattering phase function** $\Phi_\lambda(\lambda,\ \boldsymbol{v}_I,\ \mathbf{v}_S)$. This latter is defined as the ratio of monochromatic intensity scattered in direction $\mathbf{v}_S = L_S\,\boldsymbol{i} + M_S\,\boldsymbol{j} + N_S\,\boldsymbol{k}$ due to a beam incident from direction $\boldsymbol{v}_I = L_I\,\boldsymbol{i} + M_I\,\boldsymbol{j} + N_I\,\boldsymbol{k}$, to the monochromatic intensity that would be scattered in direction $\mathbf{v}_S$ if scattering were isotropic. The scattering phase function is normalized such that,

$$\frac{1}{4\pi}\int_{4\pi} \Phi_\lambda(\lambda, \vartheta_I, \varphi_I, \vartheta_S, \varphi_S) d\omega_S = 1. \tag{5.30}$$

The MCRT approach to computing the scattering phase function, while theoretically interesting, would prove to be very resource intensive in practice. It would be far more practical to use one of several online Mie scattering calculators, such as the one provided by Scott Prahl [52]. A Mie scattering Fortran program attributed to Gérard Brogniez at the

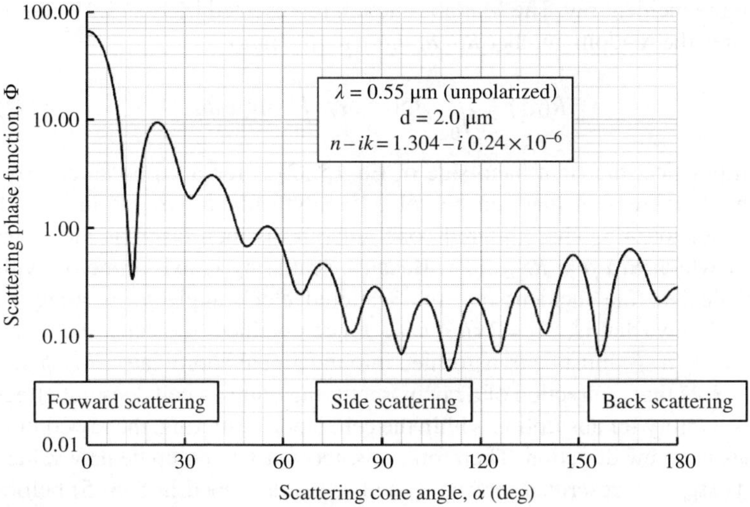

Figure 5.19 Scattering of unpolarized 0.55-μm radiation by a 2.0-μm-diameter spherical aerosol whose complex index of refraction is $1.304 - i\,0.24 \times 10^{-6}$.

Laboratoire d'Optique Atmosphérique (Université de Lille, France) [53] is available in Ref. [50]. Figure 5.19 illustrates the scattering phase function computed using the Brogniez code for unpolarized 0.55-μm radiation by a 2.0-μm-diameter spherical aerosol whose complex index of refraction is $1.304 - i0.24 \times 10^{-6}$. The scattering cone angle α in Figure 5.19 is measured with respect to the incident axis.

***Step 9*: Compute the direction of the scattered radiation.**

An energy bundle has entered a volume element with a known direction and has encountered a scattering center. In the Matlab environment the scattering model would be implemented as a suitable function of the form

$$\begin{aligned} &function[L\_S, M\_S, N\_S] \\ &\quad = \mathrm{PHI}(RS\_alpha, RS\_beta, lambda, d, n, k, L\_I, M\_I, N\_I). \end{aligned} \tag{5.31}$$

In Eq. (5.31), *L_S*, *M_S*, and *N_S* are the direction cosines of the scattered ray, *RS_alpha* and *RS_beta* are the next two available random numbers, *lambda* is the wavelength, *d* is the diameter of the scattering center, *n* and *k* are its optical constants, and *L_I*, *M_I*, and *N_I* are the direction cosines of the incident ray. The Matlab function represented by Eq. (5.31), compares the random number *RS_alpha* with the ratio

$$R(\alpha_S) = \int_0^{\alpha_S} \Phi(\alpha)d\alpha / \int_0^{\pi} \Phi(\alpha)d\alpha. \tag{5.32}$$

In practice, the right-hand side of Eq. (5.32) is replaced by an approximate expression that can be solved analytically for α_S, as in Step 1 of Section 4.5. The scattered cone angle α_S is then taken as the value for which $R(\alpha_S) = RS\_alpha$. Because scattering is axisymmetric with respect to the incident ray, the scattered clock angle β_S is assigned randomly; that is, $\beta_S = 2\pi RS\_beta$. Once obtained, the cone and clock angles (α, β) must be transformed into the global coordinates (ϑ_S, ϕ_S).

Note that redirection of a ray by scattering within a cell does not affect its liability for absorption within the cell. Upon scattering, the ray continues in a new direction. Therefore, it is necessary to compute new values of Δs_{old} (as described in Step 3) and s_s (as described in Step 5) before continuing to Step 6, as illustrated in Figure 5.20.

Emission from a surface element is a special case of emission from a volume element. The relevant source powers for each type of element are given by Eq. (5.10). For volume elements emission is assumed to be isotropic, as described in Step 2 of the current chapter; and for surface elements it is diffuse, as described in Chapter 3. Once the direction cosines have been determined, the emitted rays are treated exactly the same for both types of source.

We have now considered every possible variation of the life cycle of an energy bundle, or ray, launched into a volume element. The energy bundle eventually arrives at an absorption event, Step 8.

***Steps 8 and 8a*: Disposition of ray upon absorption.**

In these two steps we first increment a counter and check the total count of rays launched from the current starting surface or volume element

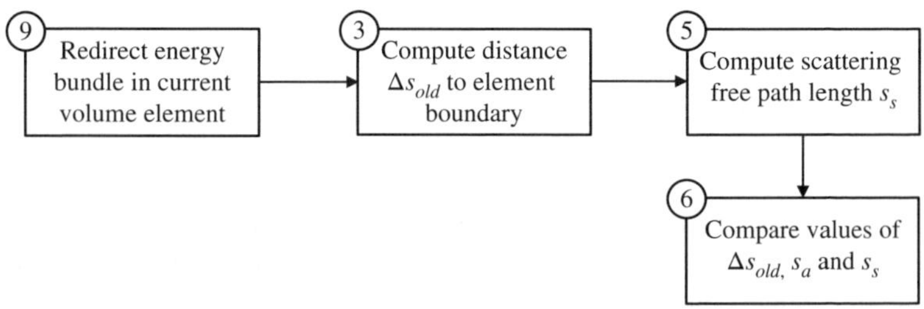

Figure 5.20 Logic flow after a scattering event.

(Step 8). Then in Step 8a we decide whether to (a) return to the current starting element and launch a new ray, or (b) terminate the ray trace for that starting element and move on to the next element. When we have repeated the process for all n elements, we are ready to estimate the radiation distribution factors.

5.4 Estimating the Radiation Distribution Factors in Participating Media

In principle, the entire operation described in this section must be repeated for each surface and volume element and for each wavelength interval $\Delta\lambda_k$ of interest. However, as in the case of the surface-to-surface distribution factors developed in Chapter 3, we will see that the number of calculations required can be halved by exploiting the reciprocity properties of the radiation distribution factor.

Recall that the radiation distribution factor D_{jik} is defined as the fraction of the power emitted from (surface or volume) element j that is absorbed in (surface or volume) element i in wavelength interval $\Delta\lambda_k$, due to both direct radiation and to all possible reflection and scattering events. This definition is the basis for Eqs. (5.8) and (5.10). Then the fraction of power emitted by surface or volume element j that is absorbed by surface or volume element i in wavelength interval $\Delta\lambda_k$ may be estimated as

$$\frac{Q_{a,jik}}{Q_{e,jk}} \equiv D_{jik} \approx \frac{N_{jik}}{N_{jk}}, \tag{5.33}$$

where N_{jk} is the number of rays emitted from surface or volume element j in wavelength interval k, and N_{jik} is the number of those rays absorbed by surface or volume element i.

As in the case of surface-to-surface radiation, we refer here to *estimating* the distribution factors rather than *computing* them. This nuance of language anticipates the material in Chapter 7, which emphasizes the statistical nature of the MCRT method. In general, the more rays that are traced, the more accurate is the distribution factor estimate. In principle, as the number of rays traced approaches infinity, the distribution factor estimate becomes exact, in which case the accuracy of the radiation heat transfer results obtained is limited only by the fidelity of the model itself. The ability to trade off accuracy requirements against the cost and availability of computing resources is one of the significant strengths of the MCRT method. More to the point, the method permits statement of the uncertainty in the heat transfer results obtained to within a specified confidence interval, as demonstrated in Chapter 7.

5.5 Using the Radiation Distribution Factors When All Temperatures are Specified

Equations (5.8)–(5.10) can be combined to yield

$$Q_i = \sum_{k=1}^{K} \left[\sum_{j=1}^{N} \varepsilon_{jk} \Delta A_j e_b(\Delta\lambda_k, T_j)(\delta_{jik} - D_{jik}) - 4\pi \sum_{j=N+1}^{n} \kappa_{jk} \Delta V_j i_b(\Delta\lambda_k, T_j) D_{jik} \right], 1 \le i \le N \tag{5.34}$$

and

$$Q_i = -\sum_{k=1}^{K} \left[\sum_{j=1}^{N} \varepsilon_{jk} \Delta A_j e_b(\Delta\lambda_k, T_j) D_{jik} - 4\pi \sum_{j=N+1}^{n} \kappa_{jk} \Delta V_j i_b(\Delta\lambda_k, T_j)(\delta_{jik} - D_{jik}) \right], \ N+1 \le i \le n \tag{5.35}$$

where

$$\delta_{jik} \equiv \begin{cases} 1, & j = i \\ 0, & j \ne i \end{cases}, \quad 1 \le k \le K. \tag{5.36}$$

It can be demonstrated that the reciprocity relation between two volume elements i and j is

$$\kappa_{ik} \Delta V_i D_{ijk} = \kappa_{jk} \Delta V_j D_{jik} \tag{5.37}$$

and the reciprocity relation between a surface element i and a volume element j is

$$\varepsilon_{ik} \Delta A_i D_{ijk} = 4\kappa_{jk} \Delta V_j D_{jik}. \tag{5.38}$$

The implications of Eqs. (5.37) and (5.38) are clear: It is only necessary to estimate half of the distribution factors using the MCRT method; the other half can be obtained using reciprocity.

Recognizing that $e_b = \pi i_b$ and introducing the above reciprocity relations, Eqs. (5.34) and (5.35) can be written

$$q_i = \sum_{k=1}^{K} \varepsilon_{ik} \left[\sum_{j=1}^{N} e_b(\Delta\lambda_k, T_j)(\delta_{ijk} - D_{ijk}) - \sum_{j=N+1}^{n} e_b(\Delta\lambda_k, T_j) D_{ijk} \right], \quad 1 \le i \le N \tag{5.39}$$

and

$$\dot{q}_i = -4 \sum_{k=1}^{K} \kappa_{ik} \left[\sum_{j=1}^{N} e_b(\Delta\lambda_k, T_j) D_{ijk} - \sum_{j=N+1}^{n} e_b(\Delta\lambda_k, T_j)(\delta_{ijk} - D_{ijk}) \right],$$

$$N + 1 \leq i \leq n. \tag{5.40}$$

Note that in Eqs. (5.39) and (5.40) the order of the subscripts on the Kronecker delta function and the radiation distribution factor are now written in alphabetical order. Note also that q_i defines a net heat flux (W m^{-2}) from a surface element, while $\dot{q}_i$ defines a net volumetric cooling rate (W m^{-3}) of a volume element. The band-averaged spectral emissivity ϵ_{ik} and the band-averaged spectral absorption coefficient κ_{ik} are averaged over the wavelength interval $\Delta\lambda_k$.

Equations (5.39) and (5.40) are explicit expressions for the net heat transfer from a given surface or volume element given the temperature distribution on the enclosure walls and within the enclosed participating medium. It is emphasized that, if all of the element temperatures are specified, there is no matrix to invert; the net heat transfers can be computed in a spreadsheet environment – simply "multiplied out" – once the band-averaged spectral distribution factors have been estimated.

5.6 Using the Radiation Distribution Factors for a Mixture of Specified Temperatures and Specified Heat Transfer Rates

Situations arise in which net heat transfer rather than temperature is prescribed for some or all of the surface or volume elements. A typical example might be the thermal design of a gas turbine combustor, where radiation combines with convection to produce one of the most challenging heat transfer analysis problems imaginable. In this situation we typically seek to predict the combustor liner temperature distribution from estimates of the volumetric heat release distribution.

The first step is to recast Eqs. (5.39) and (5.40) so that the terms representing surface and volume elements having specified net heat transfer are separated from those representing surface and volume elements having specified temperature. Suppose the first N_1 surface elements have specified net heat fluxes and unknown temperatures, and the remaining $N - N_1$ surface elements have specified temperatures and unknown net heat fluxes, while the first $n_1 - N$ volume elements have specified net cooling rates and unknown temperatures, and the remaining $n - n_1$

volume elements have specified temperatures and unknown net cooling rates. Equations (5.39) and (5.40) can then be rewritten

$$q_i = \sum_{k=1}^{K} \varepsilon_{ik} \left[\sum_{j=1}^{N_1} e_b(\Delta\lambda_k, T_j)(\delta_{ijk} - D_{ijk}) + \sum_{j=N_1+1}^{N} e_b(\Delta\lambda_k, T_j)(\delta_{ijk} - D_{ijk}) - \sum_{j=N+1}^{n_1} e_b(\Delta\lambda_k, T_j) D_{ijk} - \sum_{j=n_1+1}^{n} e_b(\Delta\lambda_k, T_j) D_{ijk} \right], \quad 1 \le i \le N \tag{5.41}$$

and

$$\dot{q}_i = -4 \sum_{k=1}^{K} \kappa_{ik} \left[\sum_{j=1}^{N_1} e_b(\Delta\lambda_k, T_j) D_{ijk} + \sum_{j=N_1+1}^{N} e_b(\Delta\lambda_k, T_j) D_{ijk} - \sum_{j=N+1}^{n_1} e_b(\Delta\lambda_k, T_j)(\delta_{ijk} - D_{ijk}) - \sum_{j=n_1+1}^{n} e_b(\Delta\lambda_k, T_j)(\delta_{ijk} - D_{ijk}) \right], \quad N+1 \le i \le n. \tag{5.42}$$

Next, we rewrite Eqs. (5.41) and (5.42) with the unknown quantities moved to the left-hand side and the known quantities moved to the right-hand side; that is,

for $1 \le i \le N_1$,

$$\sum_{k=1}^{K} \varepsilon_{ik} \left[\sum_{j=1}^{N_1} e_b(\Delta\lambda_k, T_j)(\delta_{ijk} - D_{ijk}) - \sum_{j=N+1}^{n_1} e_b(\Delta\lambda_k, T_j) D_{ijk} \right] = LHS_i$$

$$q_i - \sum_{k=1}^{K} \varepsilon_{ik} \left[\sum_{j=N_1+1}^{N} e_b(\Delta\lambda_k, T_j)(\delta_{ijk} - D_{ijk}) - \sum_{j=n_1+1}^{n} e_b(\Delta\lambda_k, T_j) D_{ijk} \right] = RHS_i; \tag{5.43}$$

for $N_1 + 1 \le i \le N$,

$$\sum_{k=1}^{K} \varepsilon_{ik} \left[\sum_{j=1}^{N_1} e_b(\Delta\lambda_k, T_j)(\delta_{ijk} - D_{ijk}) - \sum_{j=N+1}^{n_1} e_b(\Delta\lambda_k, T_j) D_{ijk} \right] - q_i = LHS_i$$

$$q_i - \sum_{k=1}^{K} \varepsilon_{ik} \left[\sum_{j=N_1+1}^{N} e_b(\Delta\lambda_k, T_j)(\delta_{ijk} - D_{ijk}) - \sum_{j=n_1+1}^{n} e_b(\Delta\lambda_k, T_j) D_{ijk} \right] = RHS_i; \tag{5.44}$$

for $N+1 \leq i \leq n_1$,

$$-4\sum_{k=1}^{K}\kappa_{ik}\left[\sum_{j=1}^{N_1}e_b(\Delta\lambda_k,T_j)D_{ijk}-\sum_{j=N+1}^{n_1}e_b(\Delta\lambda_k,T_j)(\delta_{ijk}-D_{ijk})\right]=LHS_i$$

$$\dot{q}_i+4\sum_{k=1}^{K}\kappa_{ik}\left[\sum_{j=N_1+1}^{N}e_b(\Delta\lambda_k,T_j)D_{ijk}+\sum_{j=n_1+1}^{n}e_b(\Delta\lambda_k,T_j)(\delta_{ijk}-D_{ijk})\right]$$

$$=RHS_i; \tag{5.45}$$

and for $n_1+1 \leq i \leq n$,

$$-4\sum_{k=1}^{K}\kappa_{ik}\left[\sum_{j=1}^{N_1}e_b(\Delta\lambda_k,T_j)D_{ijk}-\sum_{j=N+1}^{n_1}e_b(\Delta\lambda_k,T_j)(\delta_{ijk}-D_{ijk})\right]-\dot{q}_i$$

$$=LHS_i$$

$$4\sum_{k=1}^{K}\kappa_{ik}\left[\sum_{j=N_1+1}^{N}e_b(\Delta\lambda_k,T_j)D_{ijk}+\sum_{j=n_1+1}^{n}e_b(\Delta\lambda_k,T_j)(\delta_{ijk}-D_{ijk})\right]=RHS_i. \tag{5.46}$$

Solution of Eqs. (5.43)–(5.46) is a formidable task similar to that imposed by large parameter estimation problems. This is because the unknown temperatures and heat transfer rates are related only implicitly to the known quantities; that is, the equations are transcendental. As a solution method we recommend once again the genetic algorithm approach introduced in Chapter 4 [54–56].

5.7 Simulating Infrared Images

In 1791 Prévost [57] pointed out that a body in radiant equilibrium with its surroundings must emit and absorb heat at equal rates. It follows that all bodies at a temperature greater than absolute zero must continuously emit thermal radiation, even in the absence of other bodies. In 1847 John William Draper [58] established that radiation emitted from most solid bodies becomes visible to the human eye when the temperature of the body exceeds approximately 798 K. **Thermography** is the science of forming images of thermal radiation sources at wavelengths beyond the range of human vision, that is in the infrared. Common applications of thermography include a type of night vision goggle, infrared search-and-track systems, cameras for conducting building heat loss surveys, medical thermometry, and nighttime search-and-rescue operations, to name only a few.

Nelson [2] and Nelson et al. [59] used a thermographic camera to validate an MCRT radiant exchange model for predicting the infrared visibility of a jet engine plume. Figure 5.21 compares measured and predicted infrared jet plume images obtained during the cited study. The predicted image in Figure 5.21b was obtained using the **reverse ray-trace** method, sometimes referred to as the "line-of-sight" method [1]. It is clear that relatively few of the rays emitted from the plume can be expected to be captured by the thermographic camera. To circumvent this inefficiency, the rays are instead emitted from the camera with a directional distribution that guarantees that most of the emitted rays will intersect the plume. According to the reciprocity principle, these rays are identical to those that are emitted from the plume and intersect the camera. A virtual screen is positioned between the camera and the plume such that it

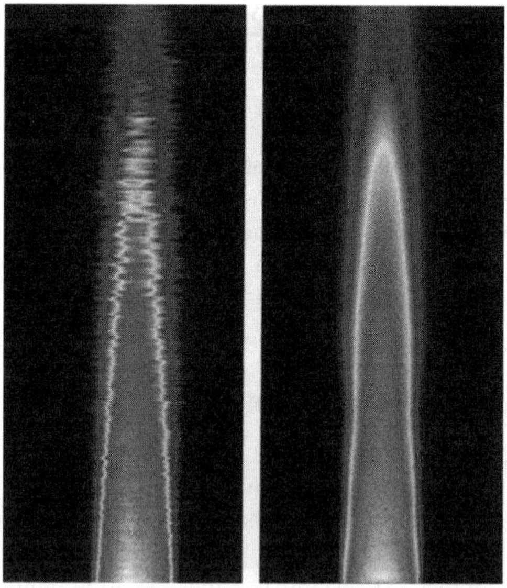

(a) (b)

Figure 5.21 Comparison of (a) measured and (b) predicted infrared images of an auxiliary power unit exhaust plume. Source: Ref. [59, reprinted with permission]. (*See color plate section for the color representation of this figure*)

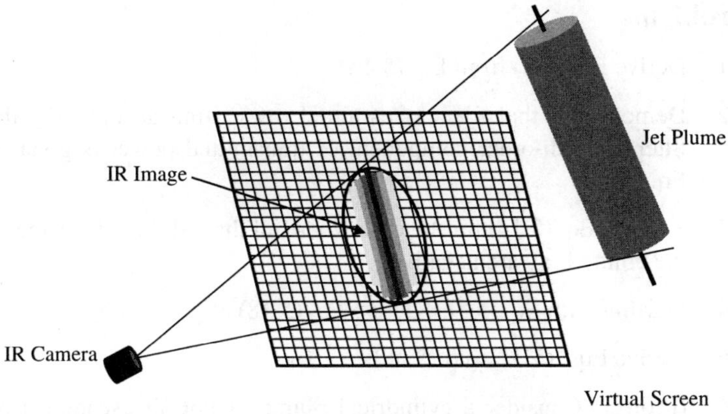

Figure 5.22 Geometry illustrating simulation of the infrared image of a jet plume.

intersects all rays passing between the camera and the plume, as illustrated in Figure 5.22. The screen is subdivided into pixels whose role is to accumulate the power of the rays passing through them.

The angular distribution of the rays leaving the camera is restricted to a cone angle ϑ_{max} as indicated in Eq. (3.16). The power assigned to each ray is

$$P_{ray} = (4\pi\kappa\Delta V)\left(\frac{\sigma T^4}{\pi}\right)\left(\frac{\Delta\omega}{4\pi}\right) = \kappa\Delta V\sigma T^4\left(\Delta A\,\frac{\cos\,\alpha}{\pi d^2}\right), \quad (5.47)$$

where κ is the absorption coefficient of the plume volume element ΔV in which the ray is absorbed, T is the local plume temperature, ΔA cos α is the projected area of the pixel with respect to the ray path, and d is the effective distance from the plume volume element to the point of intersection of the ray with the pixel. The "effective distance d" is somewhat vague, as the ray is refracted within the plume due to local temperature gradients, and, thus, consists of a series of straight-line ray segments. In principle, we should compute d as the sum of these individual segment lengths. In practice, the error associated with equating d to the straight-line distance from the center of the volume element to the point of intersection of the ray with the virtual screen would be negligible if the plume diameter is small compared to the plume-screen distance.

Problems

5.1 Derive Eq. (5.2) from Eq. (5.16).

5.2 Demonstrate that for a transmitted ray exiting an optically thin filter, the *n*th-order component of transmitted power is given by Eq. (5.6).

5.3 Rewrite Eq. (5.14) for the case where all three direction cosines of the emitted ray are negative.

5.4 Starting with Eq. (5.16), derive Eq. (5.18).

5.5 Derive Eqs. (5.37) and (5.38).

5.6 (Project) Consider a cylindrical plume of hot air escaping from a jet engine nozzle. Our task is to simulate the infrared image of a plume segment about two diameters long in which the radial temperature variation in kelvins is reasonably well approximated by $300 + 1000\,[1 - (2r/D)^2]$, where r is the radial coordinate and D = 0.25 m is the plume diameter. In the wavelength interval of interest, the absorption coefficient $\kappa_{\Delta\lambda}$ can be assumed constant at a value of $8\,\mathrm{m}^{-1}$. The refractive index of air at 300 K is 1.5. The plume should be divided into 20 equal-volume radial segments. Referring to Figure 5.22, the virtual screen should consist of a 50×50 array of pixels. Neglect scattering. Your final result should resemble the middle one-third of Figure 5.21b.

5.7 Referring to Figure 1.16 in Chapter 1, let the interior of the enclosure be filled with an emitting-absorbing medium whose absorption coefficient $\kappa_{\Delta\lambda} = 8\,\mathrm{m}^{-1}$ in the wavelength interval of interest. The enclosure dimensions are $\Delta x = \Delta y = \Delta z = 0.25$ m and the walls are diffuse-specular with $\alpha_{\Delta\lambda} = 0.55$ and $\rho^s_{\Delta\lambda} = 0.8$. Ignore the effects of refraction in this problem. Estimate all of the radiation distribution factors pertaining to this problem.

5.8 For the situation described in Problem 5.7, let Surfaces 1–4 be maintained at a temperature of 800 K and Surfaces 5–8 be maintained at a temperature of 200 K, with the remaining surfaces insulated. Compute

(a) the temperatures (K) of the insulated surfaces and
(b) the net heat transfer (W) from Surfaces 1–8.
(c) Compare these results with the results obtained for Parts (a) and (b) in Problem 3.19. Discuss any differences.

References

1. Turk, J.A., (1994). Acceleration techniques for the radiative analysis of general computational fluid dynamics solutions using reverse Monte Carlo ray tracing, PhD dissertation, Virginia Polytechnic Institute and State University.
2. Nelson, E.L, (1994). Temperature, pressure, and infrared image survey of an axisymmetric exhaust plume, PhD dissertation, Virginia Polytechnic Institute and State University.
3. Kitamura, R., Pilon, L., and Jonasz, M. (2007). Optical constants of silica glass from extreme ultraviolet to far infrared at near room temperature. *Applied Optics* 46 (33): 8118–8133.
4. Rodney, W. and Spindler, R. (1954). Index of refraction of fused quartz for ultraviolet, visible, and infrared wavelengths. *Journal of the Optical Society of America* 44: 677–679.
5. Malitson, I. (1965). Interspecimen comparison of the refractive index of fused silica. *Journal of the Optical Society of America* 55: 1205–1120.
6. Wray, J. and Neu, J. (1969). Refractive index of several glasses as a function of wavelength and temperature. *Journal of the Optical Society of America* 59: 774–776.
7. Tan, C. and Arndt, J. (2001). Refractive index, optical dispersion, and group velocity of infrared wave in silica glass. *Journal of Physics and Chemistry of Solids* 62: 1087–1092.
8. Tan, C. (1998). Determination of refractive index of silica glass for infrared wavelengths by IR spectroscopy. *Journal of Non-Crystalline Solids* 223: 158–163.
9. Tan, C. and Arndt, J. (2000). Temperature dependence of refractive index of glass SiO_2 in the infrared wavelength range. *Journal of Physics and Chemistry of Solids* 61: 1315–1320.
10. Tan, C. (1999). Optical interference and refractive index of silica glass in the infrared absorption region. *Journal of Non-Crystalline Solids* 249: 51–54.
11. Zolotarev, V. (1970). The optical constants of amorphous SiO_2 and GeO_2 in the valence band region. *Optics and Spectroscopy* 29: 34–37.
12. Popova, S., Tolstykh, T., and Vorobev, V. (1972). Optical characteristics of amorphous quartz in the 1400–200 cm^{-1} region. *Optics and Spectroscopy* 33: 444–445.
13. Girin, O., Kondratev, Y., and Raaben, E. (1970). Optical constants and spectral microcharacteristics of NaO_2-SiO_2 glasses in the IR region of the spectrum. *Optics and Spectroscopy* 29: 397–403.
14. Khashan, M. and Nassif, A. (2001). Dispersion of the optical constants of quartz and polymethyl methacrylate glasses in a wide spectral range: 0.2–3 μm. *Optics Communications* 188: 129–139.
15. Boeckner, C. (1929). A method of obtaining the optical constants of metallically reflecting substances in the infrared. *Journal of the Optical Society of America* 19: 7–15.

16. Herzberger, M. and Salzberg, C. (1962). Refractive indices of infrared optical materials and color correction of infrared lenses. *Journal of the Optical Society of America* 52: 420–427.
17. Simon, I. and McMahon, H. (1953). Study of the structure of quartz, cristobalite, and vitreous silica by reflection in infrared. *Journal of Chemical Physics* 21: 23–30.
18. Steyer, T., Day, K.L., and Huffman, R. (1974). Infrared absorption by small amorphous quartz spheres. *Applied Optics* 13: 1586–1590.
19. Yamamuro, T., Sato, S., Zenno, T. et al. (2006). Measurement of refractive indices of 20 optical materials at low temperatures. *Optical Engineering* 45 (8): 083401. 1–12.
20. Henning, T. and Mutschke, H. (1997). Low-temperature infrared properties of cosmic dust analogues. *Astronomy and Astrophysics* 327: 743–754.
21. Bach, H. and Neuroth, N. (eds.) (1998). *The Properties of Optical Glass*. Berlin: Springer Verlag.
22. Koike, C., Hasesgawa, H., Asada, N., and Komatuzaki, T. (1989). Optical constants of fine particles for the infrared region. *Monthly Notices of the Royal Astronomical Society* 239: 27–137.
23. Reitzel, J. (1955). Infrared spectra of SiO_2 from 400 cm^{-1} to 600 cm^{-1}. *Journal of Chemical Physics* 23: 2407–2409.
24. Efimov, A.M., (1990). Dispersion of optical constants of vitreous solids, PhD thesis, Vavilov State Optical Institute, Leningrad.
25. Mamedov, R.K., Mansurov, G.M., and Dubovikov, N.I. (1982). Optical constants of quartz glass in the IR range. *Optiko-Mekhanicheskaya Promyshlennost (Optical and Mechanical Industry)* 4 (56): 56.
26. Saidov, G.V. and Bernstein, E.B. (1982). Optical constants of surface layer of fused quartz in the 900–1300 cm^{-1} range. *Fizika i Khimiya Stekla* 8: 75–81.
27. Tan, G.L., Lemon, M.F., and French, R.H. (2003). Optical properties and London dispersion forces of amorphous silica determined by vacuum ultraviolet spectroscopy and spectroscopic ellipsometry. *Journal of the American Ceramic Society* 86: 1885–1892.
28. Bogens, R.K. and Zhukov, A.G. (1966). The optical constants of fused quartz in the far infrared. *Journal of Applied Spectroscopy* 25: 54–55.
29. Parker, T.J., Ford, J.E., and Chambers, W.G. (1978). The optical constants of pure fused quartz in the far-infrared. *Infrared Physics* 18: 215–219.
30. Zhilinskii, A.P., Gorchakov, A.P., Egorova, T.S., and Miskinova, N.A. (1987). Optical characteristics of fused quartz in the far IR range. *Optics and Spectroscopy* 62: 783–784.
31. Randall, C.M. and Rawcliffe, R.D. (1967). Refractive indices of germanium, silicon, and fused quartz in the far infrared. *Applied Optics* 6: 1889–1895.
32. Philipp, H.R. (1985). Silicon dioxide (SiO_2) glass. In: *Handbook of Optical Constants of Solids*, vol. I (ed. E.D. Palik), 749. New York, NY: Academic Press.
33. Miler, M. (1968). Infrared absorption of glassy silicon dioxide. *Czechoslovak Journal of Physics* 18: 354–362.

34. Ellis, E., Johnson, D.W., Breeze, A. et al. (1979). The electronic structure and optical properties of oxide glasses i. SiO_2, $Na_2O:SiO_2$ and $Na_2O:CaO:SiO_2$. *Philosophical Magazine B* 40: 105–124.
35. Beder, E., Bass, C., and Shackleford, W. (1971). Transmissivity and absorption of fused quartz between 0.2 and 3.5 μm from room temperature to 1500°C. *Journal of the American Ceramic Society* 10: 2263–2268.
36. Gillespie, D., Olsen, A., and Nichols, L. (1965). Transmittance of optical materials at high temperatures in the 1-μm to 12-μm range. *Applied Optics* 4: 1488–1493.
37. Bogdan, L. (1964). *Measurement of Radiative Heat Transfer with Thin-film Resistance Thermometers*, Contract Report NASA-27. Washington: National Aeronautics and Space Administration.
38. Sviridova, A. and Suikovskaya, N. (1967). Transparent limits of interference films of hafnium and thorium oxides in the ultraviolet region of the spectrum. *Optics and Spectroscopy* 22: 509–512.
39. Calingaert, G., Heron, S., and Stair, R. (1936). Sapphire and other new combustion-chamber window materials. *SAE Journal* 39: 448–450.
40. Heath, D. and Sacher, P. (1966). Effects of a simulated high-energy space environment on the ultraviolet transmittance of optical material between 1050 Å and 3000 Å. *Applied Optics* 5: 937–943.
41. Grenis, A.F., and Matkovich, M. J., (1965). Blackbody reference for temperature above 1200K. Study for design requirements. Technical Report AMRA-TR-65-02, Watertown, MA: Army Material Research Agency.
42. Drummond, D.G. (1936). The infra-red absorption spectra of quartz and fused silica from 1 to 7.5 μm II - experimental results. *Proceedings of the Royal Society of London. Series A, Mathematical and Physical Sciences* 153: 328–339.
43. Bagdade, W. and Stolen, R. (1968). Far infrared absorption in fused quartz and soft glass. *Journal of Physics and Chemistry of Solids* 29: 2001–2008.
44. Wong, P.T.T. and Whalley, E. (1970). Infrared and raman spectra of glasses. Part 2. Far infrared spectrum of vitreous silica in the range 100–15 cm^{-1}. *Discussions of the Faraday Society* 50: 94.
45. Vick, Brian, Mahan, J. R., and Priestley, K., Thickness allocation between two fused silica filters, 25th Conference on Characterization and Radiometric Calibration for Remote Sensing (CALCON), Space Dynamics Laboratory, Utah State University, Logan, Utah, 22–25 August, 2016.
46. Vick, B., Mahan, J.R., and Priestley, K., (2017). Complex model building: application to the earth radiation budget instrument, Proceedings of the 2nd Thermal and Fluid Engineering Conference (TFEC2017), 4th International Workshop on Heat transfer (IWHT2017), 2–5 April, Las Vegas, NV.
47. Ren, T., Modest, M.F., and Roy, S., (2017). Monte Carlo simulation for radiative transfer in a high-pressure industrial gas turbine combustion chamber, HT2017–4819, Proceedings of the ASME 2017 Heat Transfer Summer Conference, 9–12 July, Bellevue, WA.

48. Strutt, J. (Lord Rayleigh) (1871). On the scattering of light by small particles. *Philosophical Magazine, Series 4* 41: 447–454.
49. Mie, G. (1908). Contributions to the optics of turbid media, particularly of colloidal metal solutions, Translated from *Annalen der Physik* 25 (3): 377–445.
50. Mahan, J.R. (2002). *Radiation Heat Transfer: A Statistical Approach*. New York: Wiley.
51. Lenoble, J. (1993). *Atmospheric Radiative Transfer*. Hampton, VA: A Deepak Publishing.
52. Prahl, Scott, Mie scattering calculation, https://omic.org/calc/mie_calc.html (accessed 6 August 2018).
53. Brogniez, Gérard, UNO (Mie scattering phase function program), Laboratoire d'Optique Atmosphérique, Université des Sciences et Technologies de Lille, France.
54. Chambers, L. (1995). *Practical Handbook of Genetic Algorithms*. Boca Raton, FL: CRC Press.
55. Goldberg, D. (1989). *Genetic Algorithms in Search, Optimization, and Machine Learning*. New York: Addison-Wesley.
56. Davis, L. (1991). *Handbook of Genetic Algorithms*. New York: Van Nostrand Reinhold.
57. Prévost, P. (1791). Mémoire sur l'équilibre du feu. *Journal de Physique*, Paris 38: 314–322.
58. Draper, J.W. (1847). On the production of light by heat. *The London, Edinburgh, and Dublin Philosophical Magazine and Journal of Science*, Taylor & Francis 30: 345–359.
59. Nelson, E.L., Mahan, J.R., Birckelbaw, L.D., Turk, J.A., Wardwell, D.A., and Hange, C.E., Temperature, pressure, and infrared image survey of an axisymmetric heated exhaust plume. Technical Memorandum 110382, National Aeronautics and Space Administration. 1996.

6

Extension of the MCRT Method to Physical Optics

The Monte Carlo ray-trace method may be extended to describe radiant exchange for situations in which polarization, diffraction, and interference play a significant role. This requires that the definition of the "ray" be modified to include aspects of wave-like behavior such as wavelength, phase, and polarization state. Following a brief introduction to physical optics, examples from the recent literature are considered in which the MCRT method is used to evaluate the influence of these phenomena on radiation heat transfer.

6.1 Some Ideas from Physical Optics

To this point we have exploited the ray-like behavior of electromagnetic (EM) radiation to model radiation heat transfer. An alternative view of EM radiation, in which it is treated as a wave, was introduced in Chapter 2 to develop models for the interaction between EM radiation and ideal surfaces. In the current chapter we expand the traditional view of a ray to include certain wave-like phenomena. We then exploit the enhanced capability of this new ray to investigate the influence of polarization, diffraction, and interference on the performance of optical instrumentation.

The Monte Carlo Ray-Trace Method in Radiation Heat Transfer and Applied Optics, First Edition. J. Robert Mahan.

This Work is a co-publication between John Wiley & Sons Ltd and ASME Press.
Companion website: www.wiley.com/go/mahan/monte-carlo

The term "**coherence**" describes the degree to which measures of a thermal radiation field obtained at one location or at one moment in time describe the field elsewhere or at another time. Up to this point we have used the related terms "uniform" and "isotropic" to describe blackbody radiation, but we could have said that blackbody radiation is **non-coherent**. In fact, all of the thermal radiation fields encountered so far in this book are implicitly assumed to be non-coherent. We say that a monochromatic EM field is **temporally coherent** at a certain location if we can predict its phase and polarization state at an earlier or later time by observing its current phase and polarization state. Similarly, we say that the field is **spatially coherent** at a given moment in time if we can predict its phase and polarization state at another location by observing its local phase and polarization state. Furthermore, we define a **coherence time** in terms of the time period over which current knowledge of the local phase and polarization state allows meaningful prediction of those quantities at another time, and we define a **coherence length** in terms of the distance over which local knowledge of the current phase and polarization state permits a meaningful estimate of these properties elsewhere.

Natural radiation fields are non-coherent. The electric and magnetic field vectors of **naturally**, or **randomly**, **polarized** light propagating in the x-direction assume all orientations in the y,z-plane; however, they can be resolved into equal-strength **transverse-magnetic** and **transverse-electric** components. (The **linearly polarized** monochromatic wave illustrated in Figure 2.1, said to be y-polarized, is an example of a transverse-magnetic wave.) However, specular reflection from an air-water interface, passage of radiation through clouds or an aerosol-laden atmosphere, and passage through certain optical elements such as mirrors, beam splitters, and retarders ("wave-plates"), can produce wavelength-selective **partial polarization**. Partial polarization of an EM wave has the effect of increasing its coherence. The optical throughput of certain optical instrument designs is sensitive to the degree of coherence of incident light. Therefore, in some applications, accurate modeling of system optical response requires consideration of polarization.

Finally, the spatial distribution of radiation on a screen behind an aperture is wavelength dependent. Furthermore, this phenomenon, called **diffraction**, can produce **interference fringes** if coherence is sufficiently high. In some instrument designs, diffraction and interference can translate into significant measurement error.

6.2 Geometrical Versus Physical Optics

Consider the long triangular cavity, or "groove," formed by two perfectly electrically conducting planes meeting at half-angle $\vartheta = \tan^{-1}(a/2z)$, shown in cross-section in Figure 6.1. Suppose that a plane wave of monochromatic EM radiation is normally incident to the cavity opening. How is radiation reflected from the cavity? Two extremes and an intermediate situation are illustrated in the figure. Figure 6.1a depicts a situation in which the wavelength of the incident radiation is short compared to the cavity dimensions; Figure 6.1c depicts a situation in which the wavelength is long compared to the cavity dimensions; and Figure 6.1b depicts the intermediate situation in which the wavelength is on the order of the cavity dimensions.

In Figure 6.1a, the "rays" and wavefronts form a rectangular grid, with the rays extending, for all practical purposes, all the way to the surface. Essentially, all of the EM energy in the cavity is tied up in traveling waves. Because the wall is a polished metal, the rays would undergo specular reflections and eventually be reflected from the cavity. The wave view of thermal radiation comes into play only at the surface itself, as described in Sections 2.12 and 2.13. The principles of **geometrical optics** developed in Chapters 3 and 4 can be used to a high degree of accuracy to describe the situation depicted in Figure 6.1a.

In Figure 6.1b, only a relatively small number of wavelengths of the incident radiation will "fit" into the cavity; therefore, a significant fraction of the electromagnetic energy within the cavity is stored in a complex

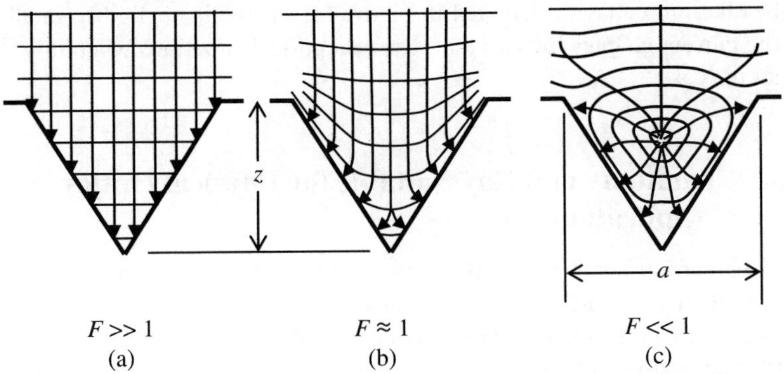

Figure 6.1 EM radiation normally incident to a long V-groove corresponding to three values of the Fresnel number, $F \equiv a\sqrt{2/\lambda z}$.

standing-wave structure. The wavefronts are distorted near the walls such that the "rays" follow curved paths there. Maxwell's equations must be solved to accurately describe the resulting potential gradients and the ultimate disposition of incident EM energy; geometrical optics can provide only an estimate of the ray pattern reflected from the cavity.

In Figure 6.1c, only a fraction of a wavelength of the incident monochromatic EM wave fits into the cavity. In this case the electrical and magnetic fields oscillate in a "bubble" structure whose complex shape is determined by the solution of Maxwell's equations subject to the zero-electric-field boundary condition imposed by the perfectly conducting walls. The contours then represent equipotential surfaces, and energy is passed back and forth between the bubble and the walls through the resulting potential gradient. The amount of reflection from the cavity would be highly dependent on the wavelength, and the directionality of the scattered radiation would be somewhat cylindrical. The principles of **physical optics** would be required to describe this situation. The **Fresnel number**, $F \equiv a\sqrt{2/\lambda z}$, where a is a characteristic width and z is a characteristic depth, can be used to separate geometrical optics ($F \gg 1$) from physical optics ($F \ll 1$), leaving a vague and broad boundary between the two. The role of the Fresnel number in physical optics is revisited in Section 6.6.

It is clear that the situation depicted in Figure 6.1c will prevail in all triangular cavities at all wavelengths sufficiently near the vertex of the cavity. This means that Planck's assumption, cited in the Preface, can never be completely valid in enclosures having corners. Fortunately, in most situations of practical engineering interest the energy involved in the aberrant behavior depicted in Figure 6.1b and c is negligibly small. The current chapter focuses on those exceptional situations when this is not the case.

6.3 Anatomy of a Ray Suitable for Physical Optics Applications

In the current chapter the MCRT method is extended to permit serious investigation of the effects of polarization and diffraction on optical instrument performance. The extended method is based on the more detailed anatomy of a ray illustrated in Figure 6.2. The two polarizations shown, s and p, do not necessarily have the same strength; that is, $|\boldsymbol{E}_p|$ may differ from $|\boldsymbol{E}_s|$. However, they are in quadrature with one another, and their mutual orientation α is established with respect to a reference,

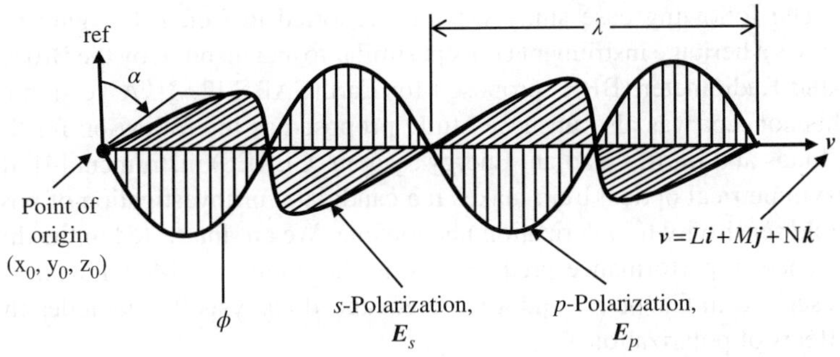

Figure 6.2 Anatomy of an individual enhanced ray.

as indicated in the figure. Thus, in addition to its point of origin (x_0, y_0, z_0) and its direction (L, M, N), the ray also must carry information about its wavelength λ, its local phase ϕ, and the orientation α and strengths $|\boldsymbol{E}_p|$ and $|\boldsymbol{E}_s|$ of its polarization. The local phase is given by

$$\phi = \phi_0 + \frac{2\pi s}{\lambda}, \tag{6.1}$$

where s is now the distance traveled from the point of origin (x_0, y_0, z_0) to an eventual point of incidence (x_1, y_1, z_1). Phase is needed for evaluation of interference when several rays are incident to the same surface element, as described in Section 6.5.

Not all of these descriptors are required for every investigation. For example, the rays used in the study of polarization effects described in Section 6.4 are not required to carry phase information, and the rays used in the study of diffraction and interference described in Section 6.5 are not required to carry polarization information.

6.4 Modeling of Polarization Effects: A Case Study

Thermal radiation emitted and reflected from the Earth and viewed from low Earth orbit is characterized by its spectral distribution, its degree of coherence, and its state of polarization. While natural radiation emitted from the sun's photosphere is randomly polarized, its reflection from oceans and cloud tops and its passage through the atmosphere produces at least partially polarized radiation at the top of the atmosphere. The possibility that the optical throughput of an Earth radiation budget instrument could be sensitive to the polarization state of incident radiation cannot be ignored.

The following case study was first reported in Ref. [1]. Figure 6.3 shows a heritage instrument concept similar to and inspired by the Broadband Radiometer (BBR) proposed for EarthCARE [2, 3]. A version of this concept was also unsuccessfully proposed as the follow-on for the Clouds and Earth's Radiant Energy System (CERES) instrument [4]. Its asymmetrical optical train makes it a candidate for investigation of possible sensitivity to polarization phenomena. We are interested in the difference in performance predicted using the traditional MCRT analysis described in Chapter 3 and a more advanced analysis that includes the effects of polarization.

Placed in low Earth orbit, the instrument shown in Figure 6.3 collects incident radiation at the top of the atmosphere and directs it through a

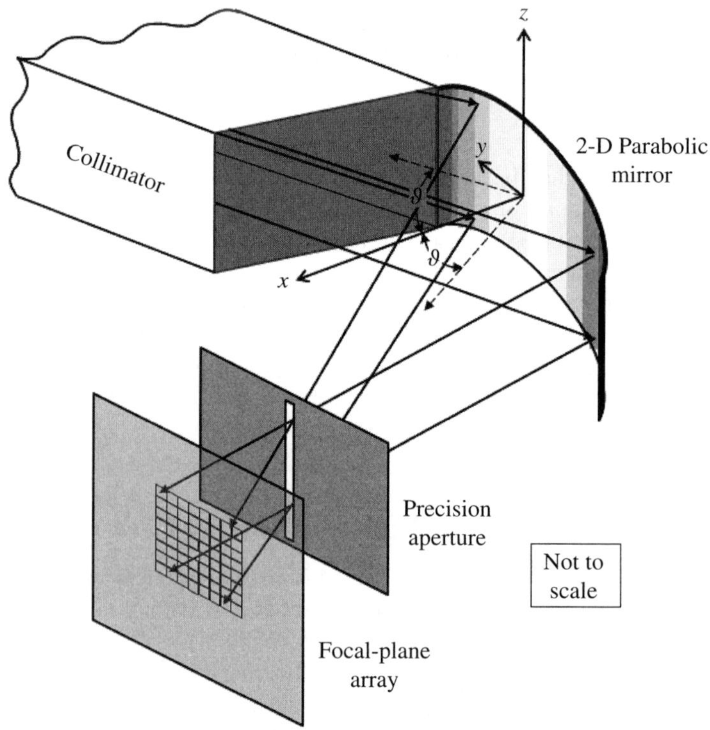

Figure 6.3 Concept for an Earth radiation budget instrument based on a two-dimensional off-axis parabolic mirror. Source: Ref. [1, reprinted with permission].

collimator onto a two-dimensional, off-axis parabolic mirror. The mirror concentrates this light onto a two-dimensional thermistor microbolometer array positioned in the focal plane. The collimator limits the angular distribution of incident radiation to an extent that depends on its length. In the current study, the width of the two-dimensional parabolic mirror is 126 cm, its height is 4.0 cm, and its focal length is 100 cm.

The collimator axis and the center of the focal plane array are both aligned at angle ϑ with respect to the optical axis of the two-dimensional parabolic mirror, as shown in the figure. This angle is a parameter of the study. In one version of the concept a precision aperture is used to define the instrument footprint on the Earth, while in another version the shape of the microbolometer array itself determines the instrument footprint. In the current section we study the effects of polarization for the version of the instrument that does not include a precision aperture. An investigation of the effects of diffraction inherent to the version that does incorporate a precision aperture is the topic of Section 6.5.

Recall from Section 6.1 that natural (non-polarized) light consists of equal amounts of power carried by each of two mutually orthogonal "polarizations," or electric field orientations. Upon incidence to a reflecting surface, each polarization can be resolved into two components: the p-polarization, whose electric field vector lies in the plane tangent to the surface at the point of incidence; and the s-polarization, whose electric field lies in the plane of incidence containing both the incident ray and the local unit normal vector. Light reflected from the ocean at the **Brewster angle** is almost entirely s-polarized. The value of the Brewster angle depends on wavelength and the local salinity of the sea water, but for visible radiation reflected from pure water it is around 53°. Clouds, molecules, and dust particles in the atmosphere can also polarize light to a degree depending on wavelength and atmospheric conditions along the path through which it propagates. Therefore, the extreme in polarization state of thermal radiation arriving at the instrument varies between s-polarized and p-polarized, with degrees of natural polarization in between.

Most if not all Earth radiation budget instruments use mirrors to focus and redirect thermal radiation onto a thermal detector such as a thermistor bolometer or a thermopile. At a given wavelength λ the reflectivity of a mirror to s- and p-polarized radiation may be well represented by Eqs. (2.91) and (2.92), where ϑ in those equations refers to the local angle of incidence to the mirror, which is generally different from the angle ϑ in Figure 6.3. The quantities n and k in Eqs. (2.91) and (2.92) are, respectively, the wavelength-dependent index of refraction and the extinction coefficient of the metallic coating forming the mirror surface. In the present investigation, the mirrors are assumed to be vapor-deposited aluminum, a common choice in Earth radiation budget instruments. Rakic's algorithm [5] for computing the optical properties of pure aluminum has been used with Eqs. (2.91) and (2.92) to determine the two extremes of polarization of the local spectral reflectivity for the two-dimensional parabolic mirror.

Two extremes of linear polarization of radiation arriving at the instrument entrance aperture are considered in this study: *p*-polarized radiation whose electric field is oriented along the z-axis in Figure 6.3, and *s*-polarized radiation. These two extremes were chosen because they produce the largest difference in instrument signal (= detector heat flux) possible between two incident beams having the same power at a given wavelength. It is also assumed that the collimator is sufficiently long to produce essentially collimated radiation arriving at angle ϑ with respect to the parabolic mirror optical axis. This assumption, too, produces the largest possible deviation between instrument responses to two beams carrying the same power at a given wavelength. Subject to these assumptions, any deviations observed in the actual instrument would be bracketed by the results obtained in this study.

The wavelength λ and the angle ϑ between the collimator axis and the mirror optical axis have been chosen as the parameters for studying the effect of polarization on instrument response. This is because the results have been found to be most sensitive to these two parameters. Also, the precision aperture has been suppressed so that the thermistor bolometer array itself, located in the focal plane, defines the Earth field-of-view. This has the effect of isolating polarization effects from any diffraction effects that might be associated with a precision aperture.

Angles of $\vartheta = 10$, 20, and 30° and wavelengths of $\lambda = 0.51$, 1.03, and 2.48 μm have been used in this study. The range of angles used is dictated by practical considerations. The best angle from optical considerations alone would be 0°, because this would avoid the image distortion, called coma, inherent to off-axis optics. However, this would require positioning the detector assembly in the path of the incident beam between the collimator and the mirror. While on-axis versions of single-mirror telescopes have been used, the resulting optical obscuration significantly reduces instrument gain. The detrimental effects of polarization and coma both increase with off-axis angle. Therefore, the instrument designer must optimize off-axis angle to avoid optical obscuration at one extreme and performance loss due to increased coma and polarization effects at the other. The wavelength interval has been chosen to span a range in which the effect of polarization varies from barely discernable at one extreme to significant at the other.

Either *s*- or *p*-polarization is attributed to each ray launched into the instrument. One million monochromatic, uniformly distributed rays exit the collimator parallel to its optical axis. Table 7.2 gives the upper bound of the uncertainty of radiation distribution factors in this case to be 0.588% within a 95% level of confidence. The wavelength λ and the angle that each ray makes with the local surface normal vector on the mirror are used with either Eq. (2.91) or Eq. (2.92), depending on the polarization of the ray, to compute the local reflectivity of the mirror.

Before exploring the effects of polarization, we first investigate the distortion of the image, called coma (see Figure 3.16), which appears as a smearing of the **point spread function** due to off-axis operation of the two-dimensional parabolic mirror. Figure 6.4 shows distributions of the number of rays, out of one million, collected in equal-area bins whose centers are distributed along a horizontal line formed by the intersection of the focal plane with the x,y-plane. The horizontal axis in these figures is $y = y' \cos \vartheta$, where y' is the projection of the y-coordinate in the focal plane. Symmetry about the x,y-plane allows the contents of the z-distributed focal-plane bins to be summed and the result consolidated at $z = 0$. The focal-plane array is the same size and has the same bin spacing in all three figures. For an off-axis angle of $\vartheta = 0$, a plot such as the ones shown in Figure 6.4 would be a single vertical line at $y = 0$ having an ordinate value of 1 000 000. The vertical axis values decrease with increasing off-axis angle because an increasing number of rays fall outside of the limits of the focal-plane array.

Figure 6.4 establishes that the **responsivity** of the device, defined as the fraction of entering rays that intersect the focal-plane array near the peak in the distribution, decreases dramatically with increasing off-axis angle. Also, the curves become less smooth moving in either direction from the peak in the distribution. This is because the sampling of the incident radiation by a given bin decreases moving away from the peak. We conclude that a linear array of microbolometers positioned parallel to the z-axis in Figure 6.3 should be positioned just to the "left-hand" side of the peak in the distribution.

Figure 6.5 shows plots of the reflectivity of *s*- and *p*-polarized rays for the range of wavelengths ($\lambda = 0.51$, 1.03, and 2.48 μm) and off-axis angles ($\vartheta = 10$, 20, and 30°) investigated. Reflectivity is used in these figures as a surrogate for the power carried by a ray. This is justified because all rays enter the instrument with the same power and are then reflected from the mirror with that power multiplied by the local spectral

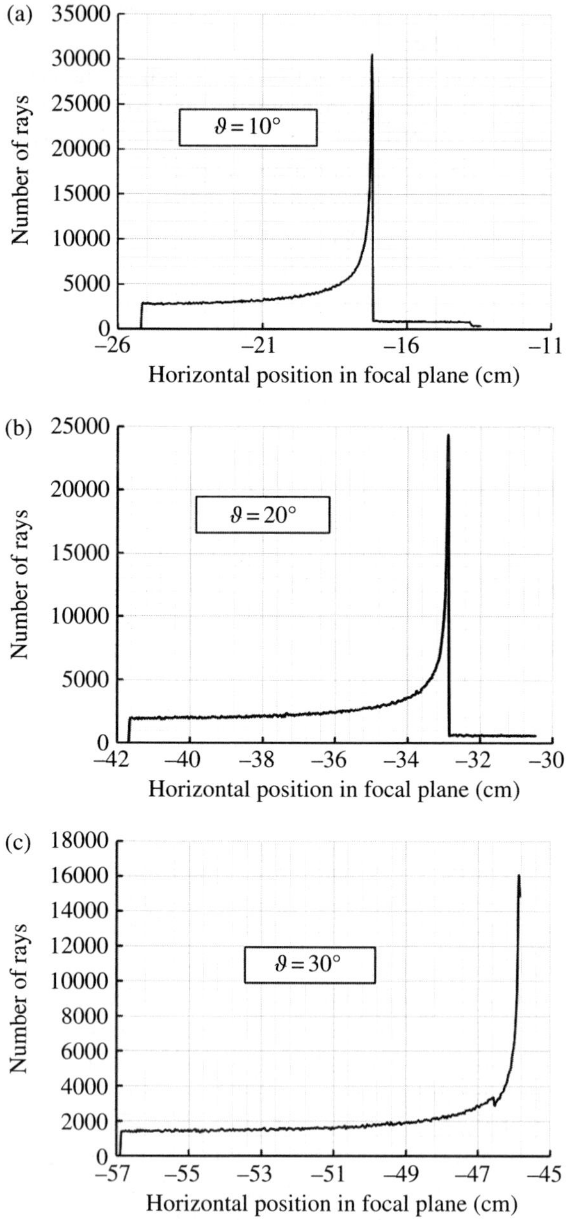

Figure 6.4 Distribution of one million rays in the focal plane for off-axis angle ϑ of (a) 10, (b) 20, and (c) 30°. Source: Ref. [1, reprinted with permission].

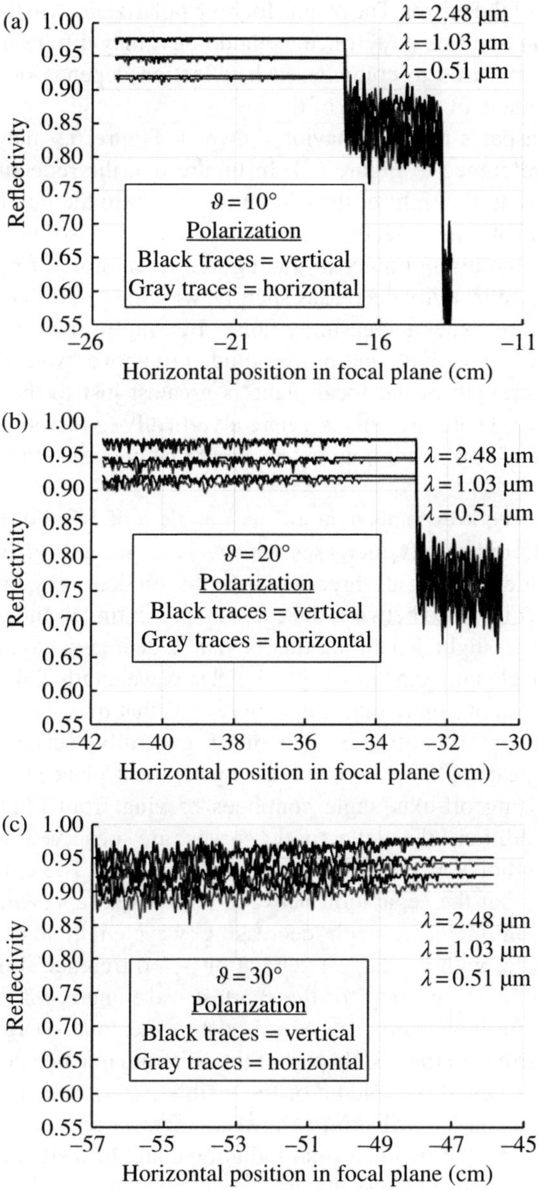

Figure 6.5 Reflectivity distributions for *s*- and *p*-polarized monochromatic rays arriving in the focal plane for off-axis angles ϑ of (a) 10, (b) 20 and (c) 30°. Source: Ref. [1, reprinted with permission].

reflectivity of the mirror. The results for both polarizations in these figures are based on traditional ray-trace methodology; they differ only because the reflectivity used to compute each reflection depends on the wavelength and angle of incidence of the ray.

Certain aspects of the behavior shown in Figure 6.5 may be interpreted by reference to Figure 6.4. In Figure 6.5, the reflectivity signal is very noisy to the right of the discontinuity where the number of rays used to estimate the reflectivity is relatively small. Also, moving to the left beyond the discontinuity in this figure, the number of rays used to estimate the reflectivity decreases sharply, with the result that the reflectivity signal becomes increasingly noisy. Bearing in mind the relationship between reflectivity and power alluded to above, we conclude that the signal strength in the focal plane is greatest just to the left of the discontinuity. Therefore, this is where a vertically (z-direction) oriented linear array of thermistor bolometers should be positioned in a practical device.

Figure 6.5a corresponds to an off-axis angle ϑ of 10°. For both polarizations, the reflectivity increases with wavelength at a given value of off-axis angle and, for all three wavelengths studied, very little separation may be discerned between the curves representing s- and p-polarized radiation. The slight separation that is visible decreases going from the shortest wavelength, 0.51 μm, to the longest wavelength, 2.48 μm.

Comparison of Figure 6.5a and 6.5b reveals that the power arriving in the focal plane for an off-axis angle of 10° generally exceeds that for an off-axis angle of 20°. This trend of decreasing focal-plane signal strength with increasing off-axis angle continues moving from Figure 6.5b to 6.5c. In addition, the reflectivity curves are observed to become increasingly noisy as the off-axis angle is increased. The critical result, however, is that the separation between the curves representing s- and p-polarization increases with decreasing wavelength and increasing off-axis angle, with the largest separation – a difference in reflectivity of about 1.1% – occurring for the shortest wavelength, 0.51 μm and at the largest off-axis angle, 30°. Now, most of the solar energy reflected from the Earth is in the visible part of the electromagnetic spectrum, and $\lambda = 0.51$ μm lies in the middle of the visible part of the spectrum. We conclude that the contribution to measurement uncertainty due to the polarization state of incident solar radiation could be a significant factor in the design of such an instrument.

6.5 Diffraction and Interference Effects: A Case Study

In the previous section the concept Earth radiation budget instrument illustrated in Figure 6.3 was operated without a precision aperture, so that the focal-plane array itself defined its footprint on the Earth. We now consider operation of the instrument with a precision aperture in place. We might conclude from the previous study that the optimum off-axis angle ϑ is 20°, so we use that angle as a basis for the study of diffraction. Furthermore, we position the precision aperture in the focal plane 2.0 mm in front of the microbolometer array. The 1.0-mm-wide precision aperture is centered at $y = -34.3174$ and $z = 0.0$ cm and extends from $y = -34.2674$ to -24.3674 cm and from $z = -2.0$ to $z = +2.0$ cm. The y-positioning of the aperture is a compromise between the need for a strong signal on the one hand and a desire to minimize the power gradient across the detector. The width of the precision aperture translates to a scientifically desirable 60-km wide footprint on the Earth for an instrument in low Earth orbit, and the distance from the precision aperture to the array minimizes the effects of beam divergence while still leaving room to insert a suitable filter.

Three wavelengths – 1.0, 10, and 34 μm – are chosen to demonstrate the influence of wavelength on diffraction in this version. Diffraction is found to be minimal at wavelengths shorter than the near infrared but begins to assert itself at longer wavelengths. Therefore, a wavelength of $\lambda = 1.0$ μm is chosen as the lower limit for the current study. The blackbody radiation distribution function for Earth-emitted radiation peaks at about 10 μm, so this wavelength is also included in the study. Finally, an exact analytical solution for diffraction at 34 μm for a similar geometry [6], verified by comparison with experimental data [7], is available for comparison with results obtained in the present study. Also, this value lies near the center of the so-called "longwave" band.

The author and his colleagues have demonstrated that diffraction in the Fresnel regime can be modeled in the MCRT environment by application of the Huygens–Fresnel principle [8]. When a ray reflected from the mirror arrives in the plane of the precision aperture, its electric field strength $E_{0,\lambda}$ is proportional to the square-root of its power; that is $E_{0,\lambda} \sim \sqrt{P_{0,\lambda}}$. The direction of the ray incident to the aperture may be resolved into a normal component, which passes through the aperture, and a tangential component, which does not. The normal component is then split into a large number – 100 000 in the present study – of subrays which are individually redirected in a fan pattern in their plane of incidence. The rule for redirecting these rays is established in Section 6.6. The distance L traveled by each subray as it passes from the precision aperture to the detector array may be converted into radians by dividing by wavelength λ and multiplying the result by 2π. Then the electric field strength E_λ at the bin is computed as

$$E_\lambda = E_{0,\lambda} \sin\left(\frac{2\pi L}{\lambda}\right). \tag{6.2}$$

These subrays are then redirected to 150 bins uniformly distributed along the width of the detector, where their electric fields are algebraically summed. At the conclusion of the ray trace, the net contribution to the electric field in each bin is squared to obtain the power absorbed in the bin. A plot of bin power versus y-position along the detector constitutes the fringe pattern produced by diffraction.

Even though a large number of rays – one million in the current study – are reflected from the mirror, only a relatively small number of these – a few thousand – are incident to the precision aperture. Similarly, many of the 100 000 subrays produced by splitting each of the rays incident to the aperture fall outside the y-range occupied by the 150-bin-wide detector array. The result is that the bin power tends to be under-sampled, leading to rather noisy-looking fringe patterns. Also, the total power level of the fringe pattern is lessened due to the rays falling outside the range of the bins.

The three graphs in Figure 6.6 show the interference fringe patterns produced in the detector plane due to diffraction by the precision aperture for wavelengths of 1.0, 10 and 34 μm. The ordinate in all three plots is the power, in arbitrary units, collected in each of the 150 bins. Recall that the power of a refracted ray has been reduced from its original value of one unit by reflection from a less-than-perfect mirror, and that only the normal component passes through the aperture. Furthermore, each subray carries only 1/100 000th of the power of a refracted ray. The "box" in the graphs whose width is that of the aperture – 1.0 mm – and whose height is 1.0 power unit represents the distribution of power on the focal plane that would result in the absence of diffraction if all of the rays arrived normal to the aperture plane with a power of one unit.

The curves in Figure 6.6 represent the fringe pattern produced corresponding to the three wavelengths for a 20° off-axis angle. The gray curve in Figure 6.6c represents an exact analytical solution [7] for the same geometry but for which the beam of incident radiation is uniform, collimated, and arrives on-axis. While there can be no expectation for the

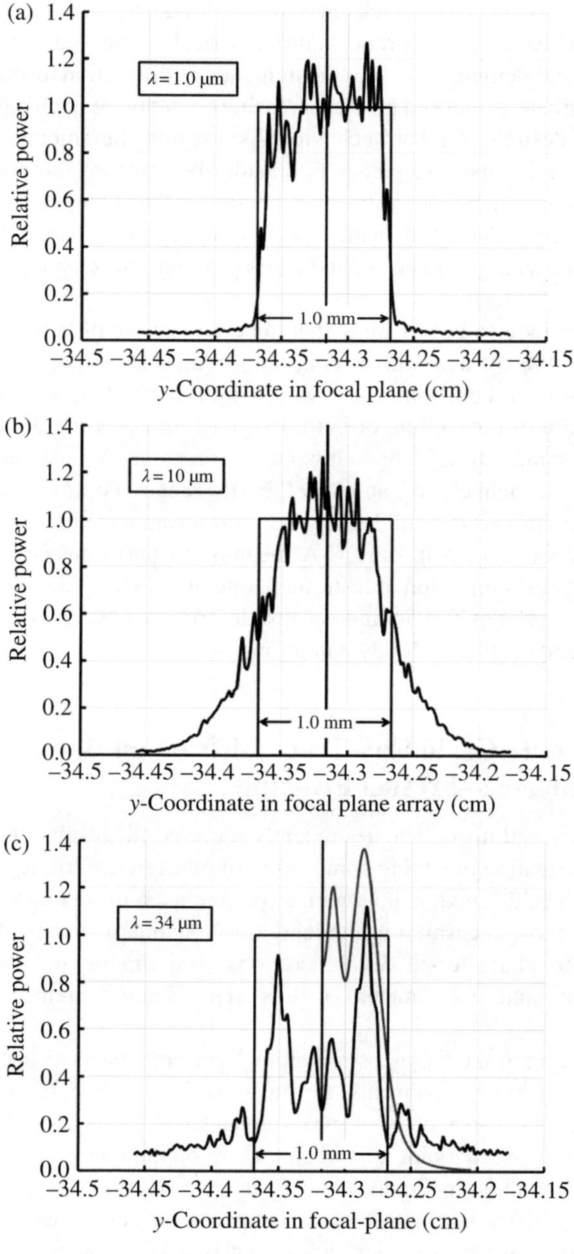

Figure 6.6 Predicted interference fringe pattern in detector plane for wavelengths λ of (a) 1.0, (b) 10, and (c) 34 μm. Source: Ref. [1, reprinted with permission].

gray curve to match the black curve in amplitude, the general agreement in shape between the theoretical and numerical results is gratifying.

Figure 6.6 demonstrates that the influence of diffraction increases with wavelength, as evidenced by the increasing deviation of the fringe pattern from the desired "top-hat" behavior. As a result, the amount of power incident to the aperture that falls outside the aperture footprint in the detector plane increases with wavelength. Furthermore, an asymmetry of the fringe pattern is evident, especially in Figures 6.6a and 6.6c. The asymmetry would be expected to be less pronounced for smaller off-axis angles.

We conclude from this study that the performance of both implementations of the off-axis mirror concept – with and without a precision aperture – would be somewhat wavelength dependent due to either polarization or diffraction, or both. However, in assessing the degree to which this might be a problem it would be necessary to take into account the spectral richness of an actual Earth scene. For example, when present, polarization would likely be more pronounced in some wavelength intervals than in others. A meaningful performance evaluation would require simulation of instrument operation when viewing a variety of Earth scene-types including oceans, deserts, forests, grasslands, snow and ice, and clear and cloudy conditions.

6.6 Monte Carlo Ray-Trace Diffraction Based on the Huygens–Fresnel Principle

The author and his colleagues recently reported [8] results of an extensive investigation involving comparison of interference fringes predicted using the MCRT method both with those predicted using classical theory and with those measured in the laboratory. Standard statistical analysis is used to characterize differences observed among the theoretical, numerical, and experimental results. This section summarizes that effort.

According to the Huygens principle, light propagates as a succession of self-replicating wavefronts, as illustrated in Figure 6.7. At its origin the wavefront for a plane wave is considered to consist of an array of equally spaced point disturbances, or monopoles, represented by the solid symbols in Figure 6.7a. Each point disturbance produces an outward-propagating pattern of concentric spherical waves. Then, for a given order of each spherical wavefront ($t + \Delta t$), a tangent plane may be passed parallel to the original plane wavefront, with each point of

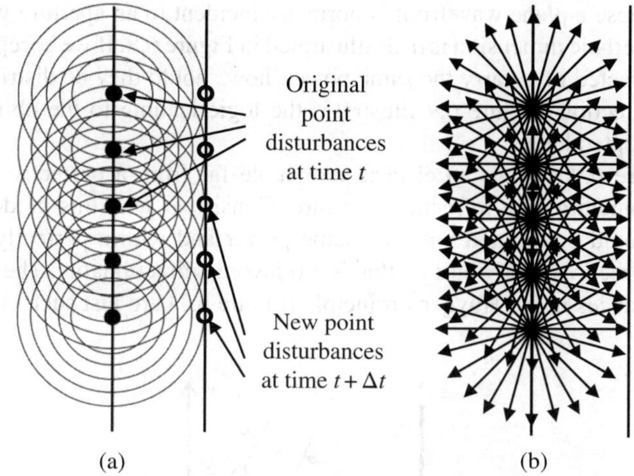

Figure 6.7 The (a) wave and (b) ray views of the Huygens principle. Source: Ref. [8, reprinted with permission].

tangency considered to be a new point disturbance, shown here as open dots. Figure 6.7b illustrates the ray view of the Huygens principle. In this view the new wavefront contains the tips of the rightward-pointing arrow heads. In this view each ray is an "enhanced" ray of the type illustrated in Figure 6.2.

The Huygens principle cannot be applied literally because at each replication of the wavefront the strength of the new generation of monopoles must be less than that of the previous generation, because half of the energy from each generation is radiated rearward. Yet we observe that the strength of a plane wave endures when propagating in a non-dissipating medium. Attempts to address this ambiguity when using the Huygens principle as the basis for formulating diffraction by apertures have led to the concept of the **obliquity factor**. Obliquity factors have been used with varying degrees of success in analytical treatments of diffraction; however, their possible role in MCRT models has been largely ignored. The traditional role of the obliquity factor is to properly model the variation of amplitude with angle ϑ for each diffracted wavelet. However, an equivalent approach, more convenient to use in the MCRT environment, is to assign the same power to all refracted wavelets but to adjust their angular density distribution to account for obliquity.

Suppose a plane wavefront is normally incident to an aperture whose characteristic dimension is a, as illustrated in Figure 6.8. If we accept that all diffracted rays carry the same power, how should they be distributed with direction? Figure 6.9 illustrates the logic leading to the obliquity rule adopted here.

Figure 6.9a shows an electric monopole far from a perfectly plane wall containing the refracting aperture. Consistent with this model, all rays would be emitted with the same power and with a perfectly uniform directional distribution; that is, no favored directionality. The point disturbances in the Huygens principle (Figure 6.7) are just such electric

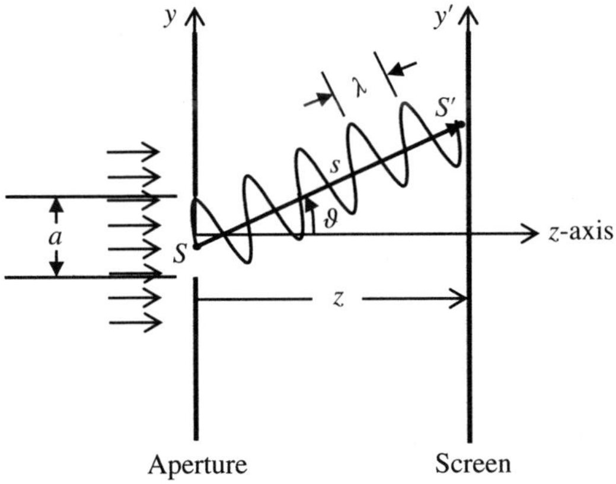

Figure 6.8 The MCRT diffraction model. Source: Ref. [8, reprinted with permission].

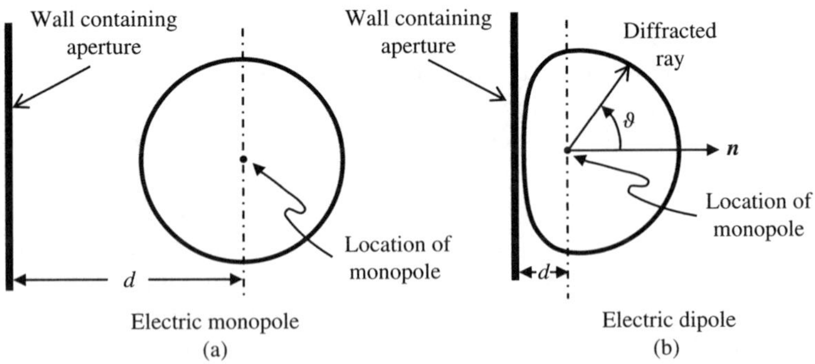

Figure 6.9 Logic for defining an obliquity rule for use in the MCRT method.

monopoles. We reject this model for use in MCRT diffraction because diffracted power flux is known to decrease with ϑ. For example, an obliquity factor attributed to Kirchhoff has the form $(1+\cos\vartheta)/2$, while the near-field analytical model [8] we use to confirm our MCRT results uses $\cos\vartheta$ [9], which corresponds to diffuse diffraction. Therefore, we use

$$\vartheta_R = \sin^{-1}\left[\sqrt{R_\vartheta}\right] \quad \text{and} \quad \varphi_R = 2\pi R_\varphi, \tag{1.32}$$

which is recognizable as the MCRT rule for diffuse reflection, where R_ϑ and R_φ are random numbers.

Figure 6.9b shows the distortion of the monopole source in Figure 6.9a that would occur if it were displaced to the left until it was only a short distance d from the wall containing the aperture. In this case the monopole source becomes a dipole source, whose directivity pattern is known to be

$$f(\vartheta) = \frac{\cos[(\pi/2)\cos\vartheta]}{\sin\vartheta}. \tag{6.3}$$

Figure 6.10 illustrates that the dipole directivity pattern $f(\vartheta)$ is nearly equivalent to the diffuse directivity pattern, $\cos\vartheta$. This leads us to speculate that the dipole directivity pattern might be physically more relevant

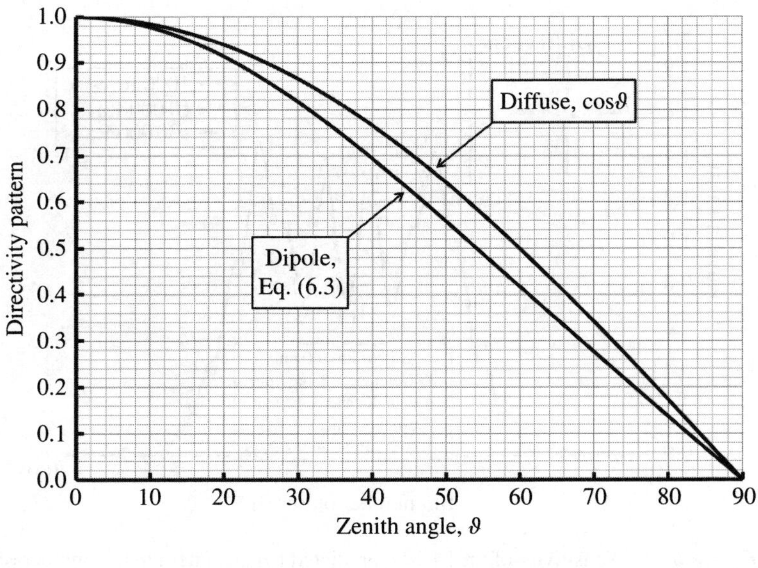

Figure 6.10 Comparison of electric dipole and diffuse directivity patterns.

to the diffraction problem while the diffuse directivity pattern is simpler and more convenient to use.

We once again invoke the Fresnel number

$$F \equiv a\sqrt{2/\lambda z} \tag{6.4}$$

first introduced in Section 6.2, where now a is the slit width or aperture diameter and z is the distance from the aperture to the screen upon which fringes are formed. Two fringe patterns produced by an aperture at the same value of Fresnel number are known to be formally similar, regardless of the values of a, λ, and z. By convention, if $F > 1.0$, diffraction is considered to be in the **Fresnel regime**, and if $F < 1.0$ diffraction is considered to be in the **Fraunhofer regime** .

Figure 6.11 illustrates the fringe pattern predicted using the MCRT method for the case of a 60-μm slit at a wavelength of 2.4462 μm and a screen distance of 100 μm, corresponding to a Fresnel number of 5.4253. Results are shown for three values of the number of rays traced from randomly located positions y in the slit: 1000, 10 000, and 100 000. A measure of the accuracy of the results obtained can be assessed by observing the departure from symmetry of the results about the center plane. The results for this case, which can be considered typical,

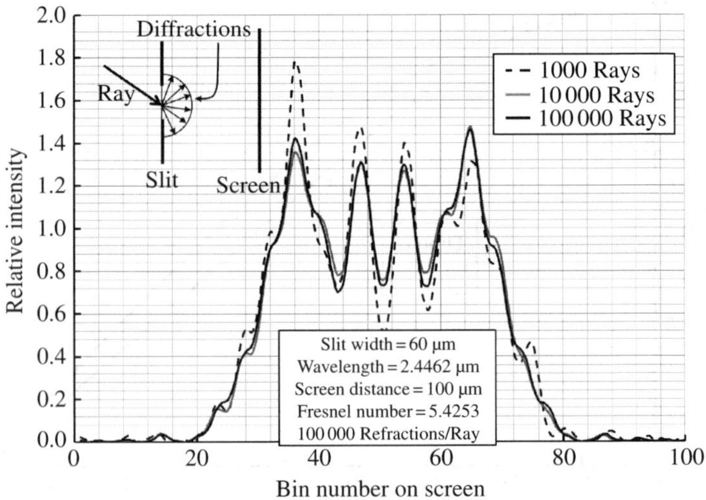

Figure 6.11 Sensitivity of the MCRT-predicted fringe pattern to the number of rays traced. Source: Ref. [8, reprinted with permission].

verify that convergence is assured when 100 000 refractions per ray are launched from 100 000 randomly located points in the slit.

To obtain data for comparison with predicted interference fringes, we measured the fringe patterns produced by a laser beam incident both to precision slits and circular apertures using the apparatus described in Ref. [8]. The MCRT results are sensitive to the number of bins into which the wavelets are bundled on the screen. The pinhole aperture of the detector used to obtain the measurements has a diameter of 2 μm, and the experimental results reported here are for slit widths and aperture diameters of 100 and 200 μm. Care was exercised to assure that the MCRT bin size roughly corresponds to the measurement spatial resolution.

The interference fringe patterns obtained in the study vary strongly with the aperture-to-screen distance z from the near-field (Fresnel) regime to the far-field (Fraunhofer) regime. The interference fringe pattern profiles produced by a slit-type aperture for six values of Fresnel number ranging from 10.18 to 0.80 are compared in Figure 6.12. For this series of images, the Fresnel number was varied by varying the aperture-to-screen distance z for a value of slit width $a = 200$ μm and a fixed wavelength of $\lambda = 351$ nm. The profiles were obtained mid-distance along a slit whose length is long compared to the slit width. The lower curves in these plots were obtained using our experimental apparatus [8] and the middle curves were obtained using standard analytical theory for the near-field Fresnel regime and the Fraunhofer approximation for the far-field regime [9]. The upper curves were obtained using the MCRT method described above. For this simulation, we used 20 000 refractions per ray launched from 20 000 randomly located points in the slit. All three curves in each set appear remarkably similar. It is particularly gratifying to see that the MCRT approach, in which the same model is used to predict fringes in both the Fresnel and Fraunhofer regimes, agrees well with the theoretical approach, in which slightly different models are used in the two regimes.

The interference fringe pattern profiles produced by a slit-type aperture for three values of the Fresnel number ranging from 6.01 to 4.03 are compared in Figure 6.13. In this series of images, a slit width a of 100 μm and the wavelength λ of 351 nm have been used. Comparison of Figures 6.12c with 6.13a, and 6.12f with 6.13c, verifies the formal similarity of fringes corresponding to approximately the same value of Fresnel number. These experimental results (lower curves) are once again compared with MCRT (upper curves) and analytical (middle curves) results.

We also measured interference fringes produced by circular apertures, and these results are compared with MCRT and analytical results in

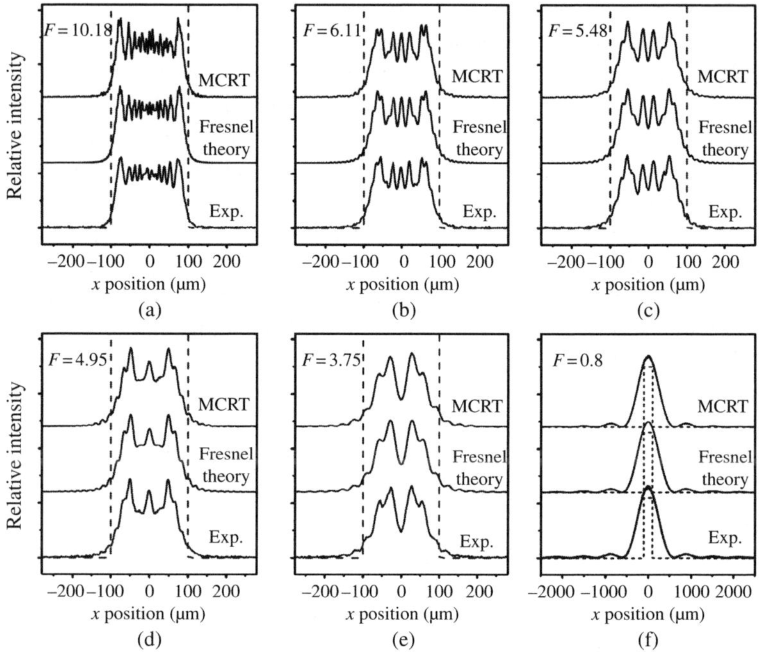

Figure 6.12 Comparison of (top to bottom) MCRT, analytical, and experimental interference fringes produced by a 200-μm slit illuminated by a 351-nm laser for a range of aperture-to-screen distances. Source: Ref. [8, reprinted with permission].

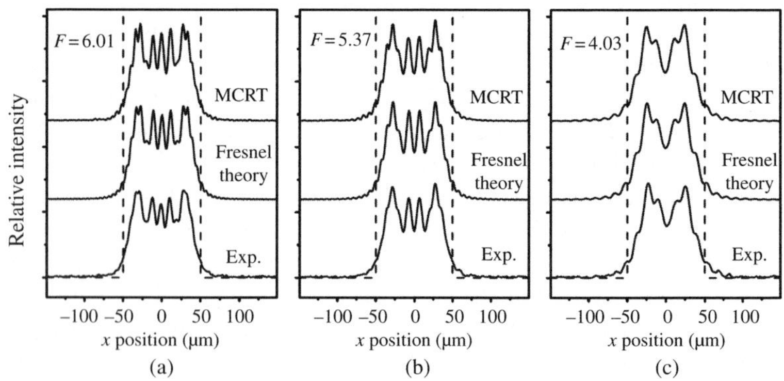

Figure 6.13 Comparison of (top to bottom) MCRT, analytical, and experimental interference fringes produced by a 100-μm slit illuminated by a 351-nm laser for a range of aperture-to-screen distances. Source: Ref. [8, reprinted with permission].

Figures 6.14–6.16. The fringes in all three figures correspond to normal illumination of the aperture by an expanded 351-nm laser beam. In Figure 6.14, the aperture diameter is 100 μm, with a screen distance of 1.9 mm; in Figure 6.15, the diameter is 200 μm with a screen distance of 7.6 mm; and in Figure 6.16 the diameter is 400 μm with a screen distance of 30.4 mm. The Fresnel number in all three figures is the same, $F = 5.48$; therefore, we expect the fringe patterns to have identical shapes even though they cover different surface areas on the screen. Intercomparison of the three figures reveals that the interference fringes are indeed formally similar even though the surface area they cover on the screen increases with increasing screen distance z, as expected. As a consequence of increasing surface area with increasing screen distance, the spatial resolution of the measured fringes (c) increases going from Figures 6.14–6.16. However, the sampling of the MCRT-based images (a) decreases as a fixed number of rays traced is spread over a larger screen area. This leads to a decreasing sharpness of the MCRT-based images moving from Figures 6.14–6.16. All three figures exhibit excellent agreement among the MCRT-based, analytical, and measured images.

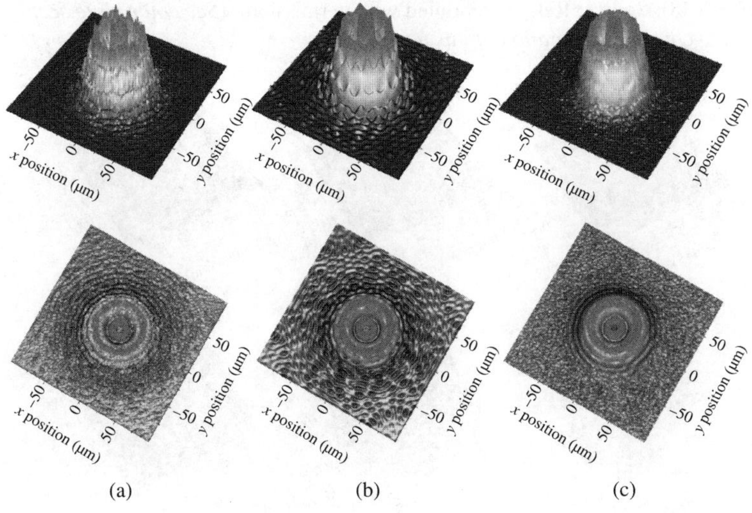

Figure 6.14 Comparison of (a) MCRT, (b) analytical, and (c) measured interference fringes corresponding to normal illumination of a 100-μm diameter circular aperture by a 351-nm laser for an aperture-to-screen distance of 1.9 mm ($F = 5.48$). Source: Ref. [8, reprinted with permission]. (*See color plate section for the color representation of this figure*)

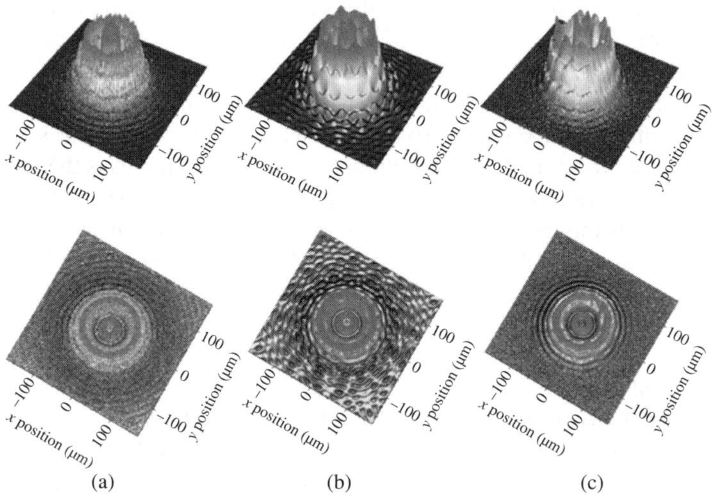

Figure 6.15 Comparison of (a) MCRT, (b) analytical, and (c) measured interference fringes corresponding to normal illumination of a 200-μm diameter circular aperture by a 351-nm laser for an aperture-to-screen distance of 7.6 mm ($F = 5.48$). Source: Ref. [8, reprinted with permission]. (*See color plate section for the color representation of this figure*)

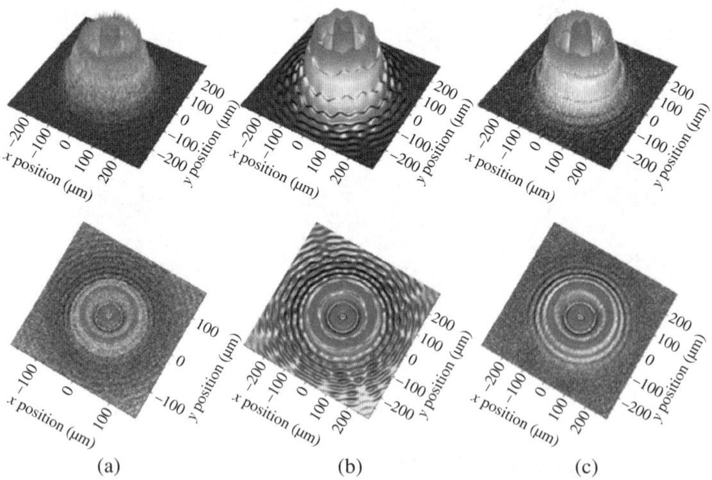

Figure 6.16 Comparison of (a) MCRT, (b) analytical, and (c) measured interference fringes corresponding to normal illumination of a 400-μm diameter circular aperture by a 351-nm laser for an aperture-to-screen distance of 30.4 mm ($F = 5.48$). Source: Ref. [8, reprinted with permission]. (*See color plate section for the color representation of this figure*)

While compelling in the examples presented here, "eyeball" comparison of results can be misleading. Therefore, three statistical approaches have been used to obtain a quantitative measure of the agreement among the three methods for obtaining interference fringes: the adjusted coefficient of determination, R^2_{adj}, the root-mean-square deviation, *RMSD*, and the reduced chi-square statistic, χ^2_ν. The most common measure is the coefficient of determination, R^2, which quantifies the **goodness of fit** of a model. The R^2 coefficient of determination,

$$R^2 = 1 - \frac{RSS}{TSS} = 1 - \frac{\sum_{i=0}^{N} (I_i^{\text{exp}} - I_i^{\text{mod}})^2}{\sum_{i=0}^{N} (I_i^{\text{exp}} - \bar{I})^2}, \tag{6.5}$$

is an estimate of the degree to which observed results are replicated by a model. In Eq. (6.5), *RSS* is the residual sum-of-squares, *TSS* is the total sum-of-squares, I_i^{exp} is the ith observed value of N observations, I_i^{mod} is the corresponding theoretical value, and $\bar{I}$ is the mean of the observed data [10]. To compensate for the possible bias that might arise due to different numbers of parameters, we employ the **adjusted coefficient of determination**, R^2_{adj}, defined

$$R^2_{\text{adj}} = 1 - \frac{N-1}{N-p-1} \times (1 - R^2), \tag{6.6}$$

where p is the number of independent variables. The adjusted coefficient of determination is always smaller than R^2.

Another view of the goodness of fit between observed and predicted results is the **root-mean-square deviation (RMSD)**,

$$RMSD = \sqrt{\frac{\sum_{i=0}^{N} (I_i^{\text{exp}} - I_i^{\text{mod}})^2}{N}}, \tag{6.7}$$

which is a measure of accuracy in predicting errors modeling a particular data set rather than between datasets.

The variance of a least-squares regression analysis is also characterized by the **chi-square statistic** [10],

$$\chi^2 = \sum_{i=0}^{N} \frac{(I_i^{\text{exp}} - I_i^{\text{mod}})^2}{\sigma_i^2}, \tag{6.8}$$

where σ_i^2 is the uncertainty in individual measurements, $I_i^{\exp}$. Finally, a refinement is the **reduced chi-square statistic**, defined

$$\chi_\nu^2 = \frac{1}{N-p} \sum_{i=0}^{N} \frac{(I_i^{\exp} - I_i^{\text{mod}})^2}{\sigma_i^2}, \tag{6.9}$$

where $\nu = N - p$ is the degrees of freedom. As a general rule, a value of reduced chi-square much greater than unity indicates a poor agreement between experimental results and the theoretical model. If the theoretical model is a good approximation, the reduced chi-square should be approximately unity, while a value of reduced chi-square much less than unity may indicate excessive pessimism about measurement errors.

Table 6.1 gives values for the adjusted coefficient of determination R_{adj}^2, the root-mean-square deviation *RMSD*, and the reduced chi-square statistic χ_ν^2, comparing both the MCRT method and the analytical method

Table 6.1 Statistical comparison of the performance of the MCRT method and the standard analytical approach for predicting measured interference fringes.

	Model	R_{adj}^2	*RMSD*	χ_ν^2
Single slit (200 μm)				
$F = 0.80$	MCRT	0.998 20	4.09×10^{-4}	1.28
	Fraunhofer approximation	0.997 56	4.79×10^{-4}	1.54
$F = 3.75$	MCRT	0.996 11	2.51×10^{-4}	0.48
	Fresnel theory	0.986 54	4.39×10^{-4}	0.36
$F = 4.95$	MCRT	0.995 31	3.01×10^{-4}	0.46
	Fresnel theory	0.994 39	3.29×10^{-4}	0.43
$F = 5.48$	MCRT	0.992 78	3.66×10^{-4}	1.88
	Fresnel theory	0.990 17	4.24×10^{-4}	0.68
$F = 6.11$	MCRT	0.990 19	4.29×10^{-4}	1.91
	Fresnel theory	0.986 80	4.97×10^{-4}	0.91
$F = 10.18$	MCRT	0.971 86	7.72×10^{-4}	1.96
	Fresnel theory	0.982 30	6.15×10^{-4}	1.34
Single slit (100 μm)				
$F = 4.03$	MCRT	0.985 49	4.28×10^{-4}	0.62
	Fresnel theory	0.992 24	3.13×10^{-4}	0.43
$F = 5.37$	MCRT	0.994 72	2.56×10^{-4}	1.05
	Fresnel theory	0.995 17	2.45×10^{-4}	0.81
$F = 6.01$	MCRT	0.987 08	4.09×10^{-4}	1.62
	Fresnel theory	0.987 03	4.09×10^{-4}	0.89

Source: Ref. [8, reprinted with permission].

with the experimental results shown in Figures 6.12 and 6.13. Inspection of the table reveals that, by all measures, both the MCRT method and the standard analytical approach are in excellent agreement with measurements of the interference fringes, and thus with each other. We conclude that the MCRT method, with the diffuse directivity model, is an effective tool for predicting the effects of diffraction and interference in optical systems.

Problems

6.1 Values of the optical constants n and k for pure aluminum according to Rakic's algorithm are provided at the companion website for this book listed on p. xix. Use these data along with Eqs. (2.91) and (2.92) to investigate the directional-hemispherical reflectivity of polished pure aluminum as a function of wavelength ($0.1 \leq \lambda \leq 3.0\,\mu\text{m}$) and incidence angle ($0 \leq \vartheta \leq 88°$). Provide and discuss appropriate plots of the results obtained. Any surprises?

6.2 Use the code created for Problem 6.1 to compute the ratio $\rho_s(\lambda, \vartheta)/\rho_p(\lambda, \vartheta)$ for polished pure aluminum at wavelengths of 0.2, 1.0, and 3.0 μm, and plot the results as a function of ϑ at each wavelength. Briefly discuss your results.

6.3 Use an MCRT-based code of your own creation to reproduce Figure 6.11.

6.4 Use the code created in Problem 6.2 to investigate the effect of angle of incidence of a collimated monochromatic beam to the aperture. Briefly discuss your results.

6.5 Using the material from this chapter, numerically simulate Young's two-slit experiment, illustrated schematically in Figure 6.17. Let P_1 and P_2 be spaced 2.0 cm apart and let the screen be 10 cm beyond the two pinholes. Let the wavelength λ of the light source be 0.5 μm, and assume that the radiation from the source is naturally polarized. Let the amplitude of the electric field at P_1 and P_2 be unity, and let the phase at P_1 and P_2 be π radians. Assume that P_1 and P_2 are two-dimensional versions of Huygens' point disturbances; that is, that they are "line" disturbances where the line is normal to the page. Therefore, they emit cylindrical rather than spherical waves. Trace a statistically significant number of rays from P_1 and P_2 to P_3 for a vertical range of 200 positions of P_3 covering ±1.0 cm. Convert the ray-trace results to relative intensity at each point on the screen.

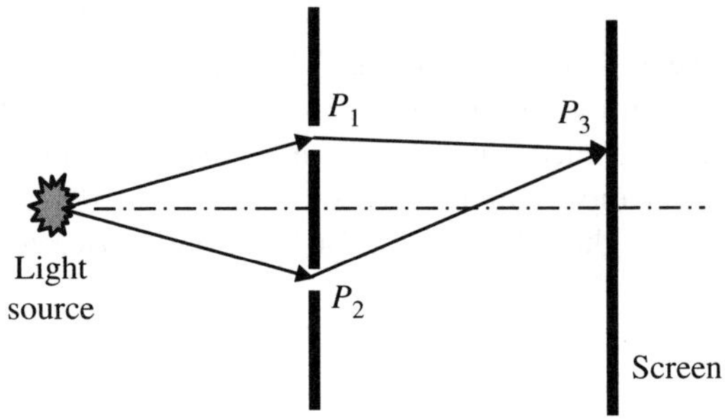

Figure 6.17 Schematic representation of Young's experiment.

(a) Execute a one-dimensional plot of the normalized intensity distribution on the screen over a vertical range of ± 1.0 cm.
(b) The "visibility" of an interference fringe pattern is defined $V \equiv (i_{\lambda,\max} - i_{\lambda,\min}) / (i_{\lambda,\max} + i_{\lambda,\min})$. What is the visibility of the interference fringes obtained in this numerical experiment?
(c) Can the visibility be improved by adjusting the pinhole-to-screen distance? Try this and report your results.
(d) Repeat Parts (a) and (b) assuming a Gaussian variation from a mean value π of the phase at P_1 and P_2 with a standard deviation of 2.0, 4.0, and 10.0%. This is equivalent to saying that the source is **quasi-monochromatic**. Compare the results obtained in Parts (b) and (d) and interpret the results in terms of **coherence**.

6.6 Go online and download the paper "A Simple Experiment on Fresnel Diffraction" by Richard E. Haskell [7]. After reading the Haskell paper, use the approach described therein to compute and display the interference fringe pattern for a wavelength of 2.4690 μm. Then use the MCRT method described in this chapter to compute the same interference fringe. Compare the results.

References

1. Mahan, J.R., Barki, A.R, Priestley, K.J., (2017). Diffraction and polarization effects in radiation heat transfer: A case study. Paper HT2017–4804, ASME 2017 Summer Heat Transfer Conference, 9–14 July, Bellevue, WA.
2. Wallace, K., Wright N., Spilling D., Ward K., and Caldwell M., (2009). The BroadBand Radiometer on the EarthCARE Spacecraft, Proceedings of SPIE 7453, 74530H, pp. 1–8.

3. Proulx, C., Allard, M., Pope, T., Tremblay, B., Williamson, F., Julien, C., Larouche, C., Delderfield, J., and Parker, D., (2010). Performance Characterization of the EarthCARE BBR Detectors. International Conference on Space Optics, Rhodes, Greece, 4–8 October.
4. Wielicki, B.A., Cess, R.D., King, M.D. et al. (November 1995). Mission to planet earth: role of clouds and radiation in climate. *Bulletin of the American Meteorological Society* 76 (11): 2125–2153.
5. Rakic, A.D. (August 1, 1995). Algorithm for the determination of intrinsic optical constants of metal films. *Applied Optics* 34 (22): 4755–4767.
6. Mahan, J.R., Barki, A.R., and Priestley, K.J. (2015). Diffraction and polarization effects in earth radiation budget measurements. *Applied Optics* 55 (34): D54–D59.
7. Haskell, R.E. (1970). A simple experiment on Fresnel diffraction. *American Journal of Physics* 38 (8): 1038–1042.
8. Mahan, J.R., Vinh, N.Q., Ho, V.X., and Munir, N.B. (2018). Monte Carlo ray-trace diffraction based on the Huygens-Fresnel principle. *Applied Optics* 57 (18): D56–D62.
9. Hecht, E. (2012). *Optics*, 5e. Pearson.
10. Bevington, P.R. and Robinson, D.K. (1992). *Data Reduction and Error Analysis for the Physical Sciences*, 2e. New York: McGraw-Hill.

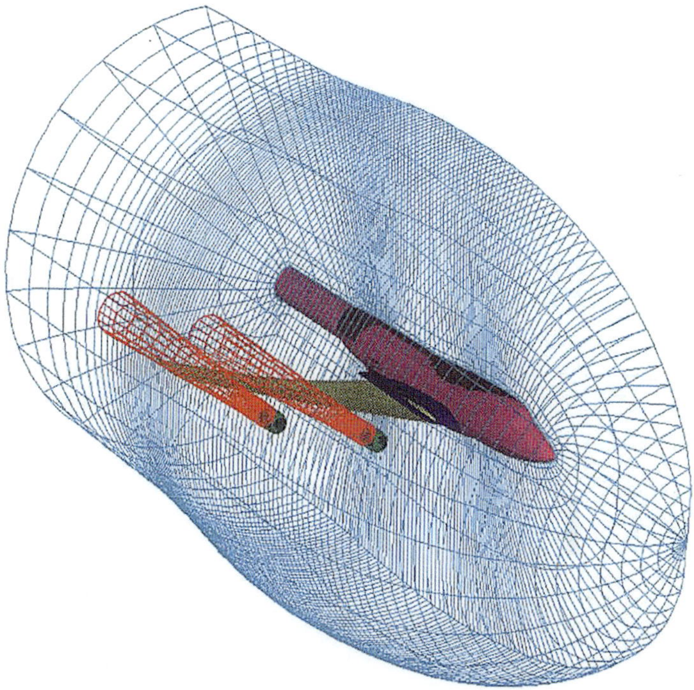

Figure 5.1 Overlay of several structured computational fluid dynamics (CFD) meshes of a Boeing 747. Source: Ref. [1, reprinted with permission].

The Monte Carlo Ray-Trace Method in Radiation Heat Transfer and Applied Optics,
First Edition. J. Robert Mahan.

This Work is a co-publication between John Wiley & Sons Ltd and ASME Press.
Companion website: www.wiley.com/go/mahan/monte-carlo

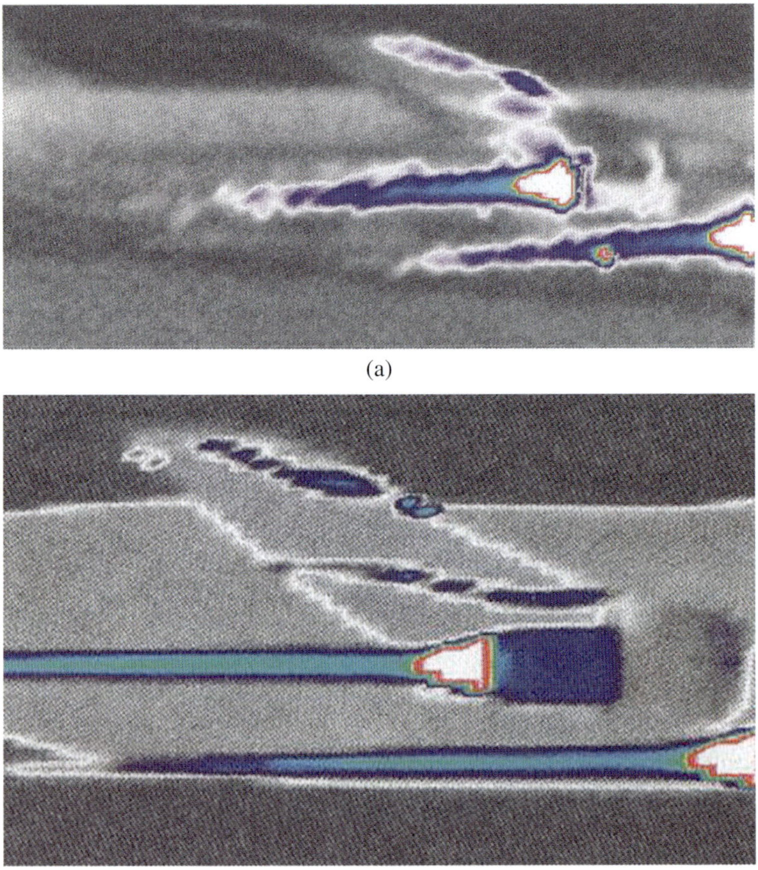

(a)

(b)

Figure 5.2 Comparison of (a) measured and (b) predicted mid-wavelength range infrared images of a Boeing 747 in flight. Source: Ref. [2, reprinted with permission].

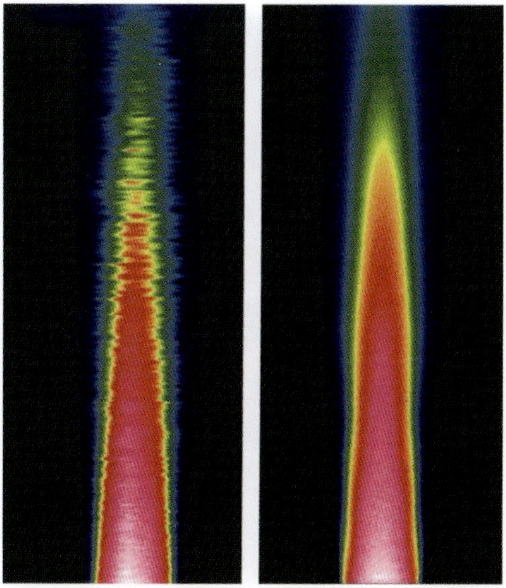

(a) (b)

Figure 5.21 Comparison of (a) measured and (b) predicted infrared images of an auxiliary power unit exhaust plume. Source: Ref. [59, reprinted with permission].

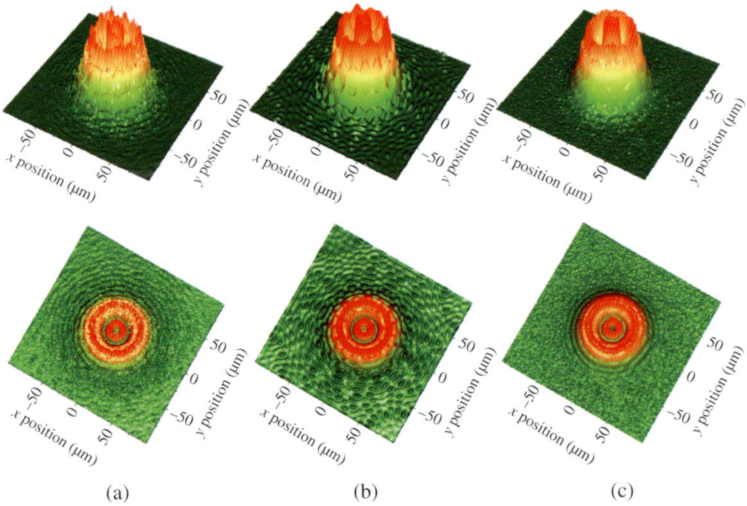

Figure 6.14 Comparison of (a) MCRT, (b) analytical, and (c) measured interference fringes corresponding to normal illumination of a 100-μm diameter circular aperture by a 351-nm laser for an aperture-to-screen distance of 1.9 mm ($F = 5.48$). Source: Ref. [8, reprinted with permission].

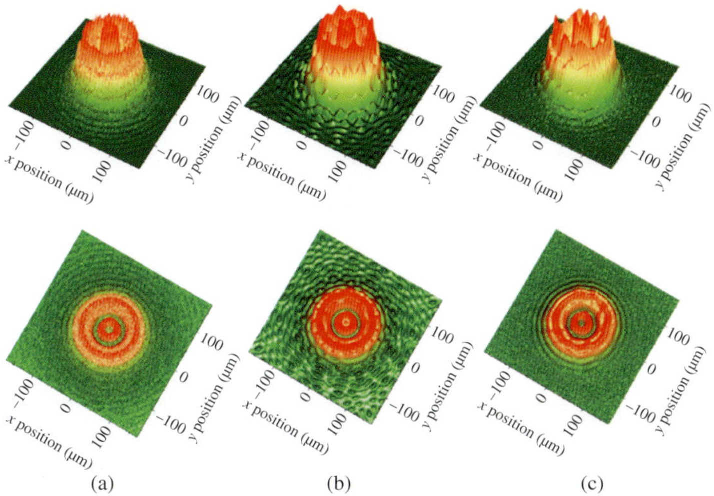

Figure 6.15 Comparison of (a) MCRT, (b) analytical, and (c) measured interference fringes corresponding to normal illumination of a 200-μm diameter circular aperture by a 351-nm laser for an aperture-to-screen distance of 7.6 mm ($F = 5.48$). Source: Ref. [8, reprinted with permission].

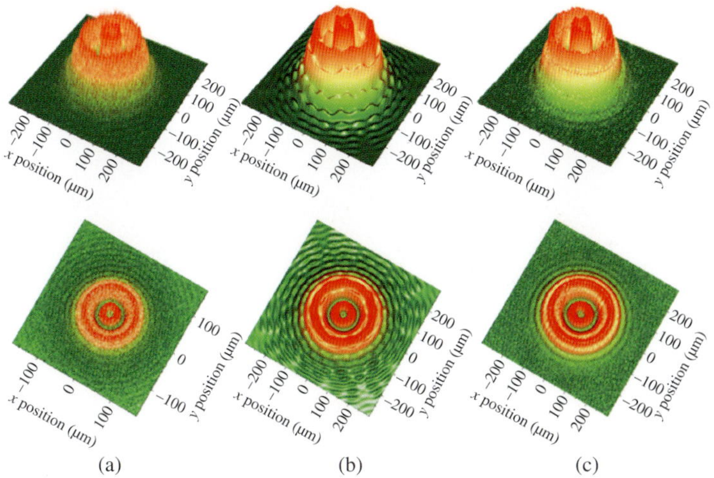

Figure 6.16 Comparison of (a) MCRT, (b) analytical, and (c) measured interference fringes corresponding to normal illumination of a 400-μm diameter circular aperture by a 351-nm laser for an aperture-to-screen distance of 30.4 mm ($F = 5.48$). Source: Ref. [8, reprinted with permission].

7

Statistical Estimation of Uncertainty in the MCRT Method

The Monte Carlo ray-trace (MCRT) method is based on a probabilistic interpretation of the radiative behavior of surface and volume elements, and the radiation distribution factor is itself a probability. Therefore, the uncertainty of results obtained using the method should be predictable using standard statistical methods. Specifically, we should be able to use statistical inference to state, to a specified level of confidence, the uncertainty of a result obtained. The chapter begins with a brief review of probability and statistics, after which the principles of statistical inference are applied to the MCRT method. Finally, a formal structure is presented for the experimental design of MCRT algorithms.

7.1 Statement of the Problem

The **uncertainty** in the value of a physical quantity is generally taken to be the difference between the true and observed values of the quantity, where the true value is unknown. A procedure is sought that provides an

The Monte Carlo Ray-Trace Method in Radiation Heat Transfer and Applied Optics, First Edition. J. Robert Mahan.

This Work is a co-publication between John Wiley & Sons Ltd and ASME Press.
Companion website: www.wiley.com/go/mahan/monte-carlo

estimate, with a stated level of confidence, of the uncertainty of a result obtained using the Monte Carlo ray-trace method in radiation heat transfer analysis.

The problem of estimating the uncertainty of a result obtained using the MCRT method consists of three steps:

1. Quantify the uncertainty of the estimates of the distribution factors.
2. Quantify the uncertainty of the model that uses the distribution factors, assuming that the uncertainty in the distribution factors is known.
3. Combine the two components of the uncertainty in a statistically valid manner.

7.2 Statistical Inference

The MCRT method allows us to obtain an **estimate** of the distribution factor based on a **sample** of a large **population**. In principle, the "true" value of the distribution factor is based on an infinite population of uniformly distributed random numbers, while the estimate is based on a large but finite sample of this population. The validity of the method is based on the hypothesis that the statistical behavior of the sample reflects that of the population from which it is drawn. This also turns out to be one of the underlying hypotheses of **statistical inference**. Therefore, we are encouraged to seek a means of applying statistical inference to results obtained using the MCRT method.

The distribution of a population is characterized by a **mean** μ and **standard deviation** σ, neither of which is known in most applications of practical interest. A third property of a population distribution is the **proportion** π having a certain characteristic. The proportion property divides a distribution into two parts (male and female, old and young, heads and tails, and so forth). It is convenient to think of the total number of energy bundles emitted from surface or volume element i as being divided into two groups: those absorbed by surface or volume element j, and those *not* absorbed by surface or volume element j. The true value of the distribution factor D_{ij}^t (which is never known) is then the **probability** that an energy bundle emitted by a given surface or volume element will be absorbed by some other specified surface or volume element. Because they are probabilities, all of the distribution factors from a given surface or volume element sum to unity over the enclosure. Thus, if D_{ij}^t is the probability that an energy bundle emitted from surface or volume element i will be absorbed by surface or volume element j, then

$1 - D^t_{ij}$ is the probability that the energy will be absorbed elsewhere in the enclosure. Thus, we have the pair of possible results "absorbed by *j*" and "not absorbed by *j*." It is in this sense that the radiation distribution factor can be thought of as a proportion.

The **sample mean** m, the **sample standard deviation** s, and the **sample proportion** p are **estimators** of, respectively, the population mean μ, the population standard deviation σ, and the population proportion π. Statistical inference in the present context consists of using a sample of the population to obtain values of m, s, and p whose deviations from μ, σ, and π are bounded with a known level of confidence. The validity of statistical inference increases with the size of the sample as long as the sample remains small compared to the population from which it is drawn. This is because the statistical distribution of the estimators becomes increasingly normal under these sample size restrictions, which clearly apply in the MCRT environment.

In the realm of statistics the term **confidence** bears a formal relationship to the probability that the value of a parameter will fall within a stated interval. The **degree of confidence** is normally stated in terms of a **confidence interval**. For example, a 95% confidence interval means that the probability of the true value of a parameter lying within a specified interval is at least 0.95. (Formally, statisticians prefer to say that the proposition that the true value lies within the specified range holds with 95% confidence.)

The relevant statistical tool for computing confidence intervals for the mean μ is **Student's *t* statistic**,

$$t = \frac{m - \mu}{s/\sqrt{N}}, \tag{7.1}$$

where the sample standard deviation s is

$$s = \sqrt{\frac{1}{N-1}\sum_{i=1}^{N}(x_i - m)^2}, \tag{7.2}$$

and N is the number of observations in the sample. If the sample size is sufficiently large (but still small compared to the population size), Student's t statistic is related to the **normal probability distribution function**,

$$p(z) = \frac{1}{\sigma\sqrt{2\pi}} e^{-\frac{(z-\mu)^2}{2\sigma^2}}, \tag{7.3}$$

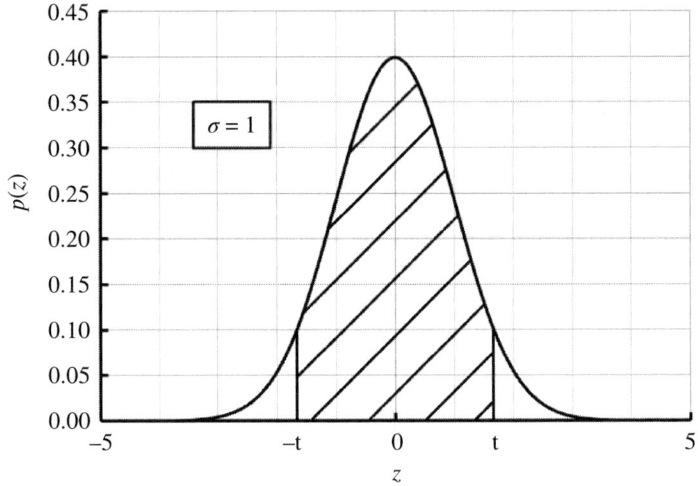

Figure 7.1 The normal probability distribution function showing the values of *t* corresponding to a given confidence interval.

as indicated in Figure 7.1. The total area under the curve is unity, and the shaded area under the curve, bounded by $\pm t$, is the probability $P(t)$. In the context of Eq. (7.1), $P(t)$ is interpreted as the level of confidence that the true mean μ lies on the interval

$$m - t\frac{s}{\sqrt{N}} \leq \mu \leq m + t\frac{s}{\sqrt{N}}. \tag{7.4}$$

Example Problem 7.1

A certain pseudo-random number generator (PRNG) is claimed to provide random numbers uniformly distributed between zero and unity. We would like to know with 95% confidence the upper and lower limits on the mean value of the random numbers provided by this PRNG. We will base our estimate on the limited sample of 35 random numbers whose values are plotted in Figure 7.2. The mean value of the sample is found to be $m = 0.4405$ with a sample standard deviation $s = 0.2970$. The value of the *t* statistic as a function of confidence interval is commonly tabulated in statistics texts. For a 95% confidence interval with a sufficiently large population ($N > 30$) we find that the critical value of the Student's *t* statis-

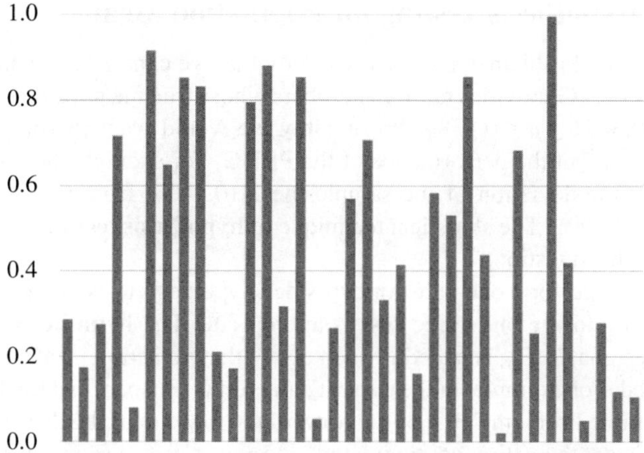

Figure 7.2 A sample of 35 pseudo-random numbers provided by a PRNG.

tic is $t_c = 1.960$. Then with a 95% level of confidence we can state that the true mean of the random numbers provided by the candidate PRNG lies in the interval

$$m - 1.960\frac{s}{\sqrt{N}} \leq \mu \leq m + 1.960\frac{s}{\sqrt{N}}, \tag{7.5}$$

or

$$0.4405 - 1.960\frac{0.2970}{\sqrt{35}} \leq \mu \leq 0.4405 + 1.960\frac{0.2970}{\sqrt{35}}, \tag{7.6}$$

or finally

$$0.3421 \leq \mu \leq 0.5389. \tag{7.7}$$

This is not really a very satisfying result because (i) the sample mean (0.4405) deviates significantly from its expected value (0.5000), and (ii) the range of values where we can be 95% confident of finding the true mean of the PRNG is fairly broad. Based on this analysis we might be tempted to conclude that our PRNG is biased toward the low end of its range, that is toward zero. However, that might be a hasty conclusion in the sometimes surprising world of statistics, as we shall see in Example Problem 7.2.

7.3 Hypothesis Testing for Population Means

In Example Problem 7.1 we learned only that we can be 95% confident that the PRNG provides random numbers whose true mean value is in the range $0.3421 \leq \mu \leq 0.5389$. Presumably we would want to know more than this about the performance of the PRNG. For example, how **significant** is the deviation of the sample mean (0.4405) from the expected value (0.5000)? The statistical technique of **hypothesis testing** is used to answer this question.

We may perform one-sided or two-sided hypothesis tests. In the current scenario a **lower one-sided test** is appropriate. The **hypothesis under test**, designated H_0, is stated in such a way that no action is taken if it is accepted. For example, in the current situation we suspect that the PRNG of Example Problem 7.1 is biased toward the low end of the distribution. If this is so, the action contemplated is to reject it for use in our MCRT studies. Therefore, the hypothesis under test is

$$H_{0:}\,\mu \geq 0.5000. \tag{7.8}$$

If the test fails, the hypothesis is rejected and we accept instead the **alternative hypothesis**,

$$H_{1:}\,\mu < 0.5000, \tag{7.9}$$

and conclude that, to a stated level of significance, the PRNG is biased toward the lower limit. However, if the test is passed we accept H_0 and conclude that, to the stated level of significance, the PRNG is not biased toward the lower end of the range.

Two levels of error can occur in hypothesis testing [1]:

1. Type I error in which we reject H_0 when it is in fact true.
2. Type II error in which we accept H_0 when H_1 is true.

The type I error – rejecting a true hypothesis – is generally considered more serious than the type II error – accepting a false hypothesis. For this reason, the test is designed to minimize the incidence of type I errors, that is, to minimize taking action when no action is called for.

The **significance level** α is the upper bound on the probability of rejecting H_0 when it is, in fact, true. Then we set out to test H_0 to a stated level of significance, say 5% ($\alpha = 0.05$). The basis of our test is once again Student's t statistic. We use Eq. (7.1) to compute the value of t. We then reject H_0 if t is less than or equal to (in the algebraic sense) a **critical value**

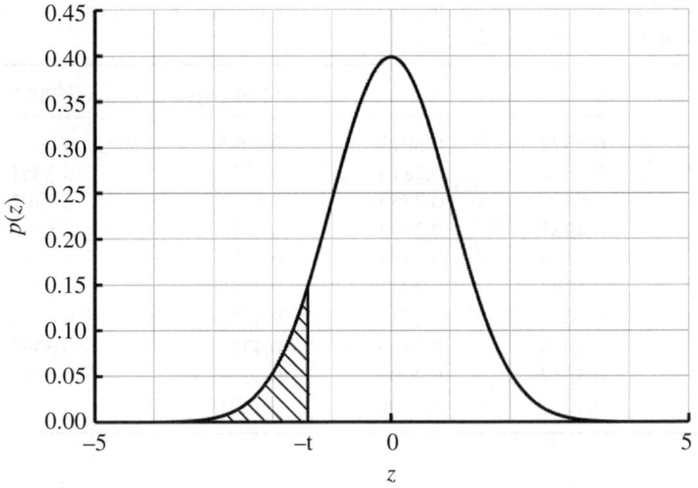

Figure 7.3 The normal probability distribution function showing the critical value of t for a one-sided hypothesis test.

based on the normal probability distribution function shown in Figure 7.3. Specifically, the critical value of t (<0 in this example) delineates an area (shown shaded in Figure 7.3) under the normal probability distribution function curve, between $-\infty$ and t, equal to α.

Example Problem 7.2

Is the PRNG of Example Problem 7.1 biased toward the low end? To answer this question we test the hypothesis

$$H_0: \mu \geq 0.5000. \tag{7.10}$$

The values for the sample mean m (=0.4405), the expected μ (=0.5000), the sample standard deviation s (=0.2970), and the sample size N (=35) are introduced into Eq. (7.1). When this is done there results $t = -1.1848 > t_c = -1.645$, where the critical value is read from a standard table. Therefore, we accept the hypothesis under test and conclude that the PRNG is unbiased to a 5% level of significance. This significance level is usually associated with "ordinary" concern about making a type I error, while a 1% significance level is associated with "extreme" concern and a 10% significance level with "mild" concern.

Table 7.1 Summary of results obtained using the PRNG to obtain 10 sets of 350 random numbers each.

i	m_i	s_i	Lower limit	Upper limit
1	0.5322	0.3011	0.5006	0.5637
2	0.5056	0.2813	0.4762	0.5351
3	0.5419	0.2753	0.5131	0.5707
4	0.4967	0.2792	0.4675	0.5260
5	0.5139	0.2884	0.4837	0.5441
6	0.4954	0.2861	0.4654	0.5254
7	0.5042	0.2784	0.4750	0.5333
8	0.5036	0.2966	0.4725	0.5347
9	0.5077	0.2904	0.4773	0.5381
10	0.4975	0.2908	0.4671	0.5280

It should be noted here that a larger sample size than $N = 35$ in Example Problems 7.1 and 7.2 would most likely yield a value of m closer to 0.5000, a smaller sample standard deviation, and a narrower 95% confidence interval. Ten experiments were conducted in which 350 random numbers were drawn from the same PRNG in each experiment. The results are summarized in Table 7.1. The mean of the mean,

$$\langle m \rangle = \frac{1}{10} \sum_{i=1}^{10} m_i, \tag{7.11}$$

for the data given in Table 7.1 is 0.5099. The upper and lower limits listed in Table 7.1 are computed using Eq. (7.5). These results, based on 3500 random numbers, support the conclusion of the hypothesis test, based on a sample of only 35 random numbers, that the PRNG is not biased toward the low end. It should be remarked that 3500 is still a very small number of random numbers in the context of a MCRT analysis.

7.4 Confidence Intervals for Population Proportions

It has already been established that the true value of the distribution factor D_{ij}^t is a **population proportion**. We would like to know, to a stated level of confidence, the maximum width of the interval containing the true value of the distribution factor (corresponding to an infinite number of energy bundles traced) and the estimate obtained using a large but finite number of energy bundles.

Recall that population proportions come in pairs (heads and tails, male and female, etc.). Such paired variables are often referred to as **Bernoulli variables**. Population proportions follow a **binomial probability distribution**,

$$P(k) = \binom{N}{k} \pi^k (1 - \pi)^{N-k}, \tag{7.12}$$

where

$$\binom{N}{k} = \frac{N!}{k!(N-k)!}. \tag{7.13}$$

In Eq. (7.12), π is the population proportion (a number between zero and unity), and in Eqs. (7.12) and (7.13) N is the number of observations, or "experiments," and k is the number of distinguishable ways a particular outcome of the experiment can occur. As an example, what is the probability that two "fair" coins, when tossed on the floor, will end up one heads and one tails? In this example $\pi = 0.5$ (equal chance of landing heads up), $N = 2$ (two independent coins), and $k = 1$ (H|T is indistinguishable from T|H). The probability of getting one heads and one tails (in no particular order) is then

$$P(H \mid T \text{ or } T \mid H) = \left(\frac{2!}{1!1!}\right)(0.5)^1(1 - 0.5)^1 = \frac{1}{2}. \tag{7.14}$$

Now if a population is **randomly** sampled for proportion and the sample size is sufficiently large, the statistical distribution of the sample proportion p will be normal even though the underlying distribution is binomial. This means that hypothesis testing for proportion can be done using essentially the same statistical tools as used in the case of the population mean. The only difference is that now our estimate of the sample standard deviation is based on the standard deviation of a binomial distribution; that is,

$$\sigma = \frac{\sigma_p}{N}, \tag{7.15}$$

where

$$\sigma_p \equiv \sqrt{N\pi(1 - \pi)}. \tag{7.16}$$

The estimator of the standard deviation for a binomial distribution in Eq. (7.16) is

$$s_p = \sqrt{Np(1 - p)}, \tag{7.17}$$

in which case

$$s = \sqrt{\frac{p(1-p)}{N}}. \tag{7.18}$$

The equivalent of the t statistic in establishing population proportion confidence intervals is the W statistic,

$$W = \frac{p - \pi}{\sqrt{p(1-p)/N}}, \tag{7.19}$$

where the probability distribution of W is approximately normal. Then the confidence interval for the population proportion π is

$$p - W_c\sqrt{\frac{p(1-p)}{N}} \leq \pi \leq p + W_c\sqrt{\frac{p(1-p)}{N}}, \tag{7.20}$$

where the critical value of W, W_c, is determined exactly as in the case of population mean. So, for example, for a 95% confidence interval with $N > 30$, $W_c = 1.960$.

Example Problem 7.3

An MCRT study based on the emission of 100 000 energy bundles leads to an estimate for the total distribution factor from surface i to surface j of $D_{ij}^e = 0.00263$ (that is 263 of the 100 000 energy bundles emitted from surface element i were absorbed by surface element j). Compute the 95% confidence interval for this result.

We use Eq. (7.19) with $p = 0.00263$, $W_c = 1.960$, and $N = 100\,000$ and obtain

$$0.00231 \leq D_{ij}^t \leq 0.00295.$$

We conclude with a 95% level of confidence that the true value of the distribution factor (obtained by tracing an infinite number of energy bundles) lies between 0.00231 and 0.00295. That is, to a 95% level of confidence,

$$D_{ij}^t = 0.00263 \pm 0.00032, \quad \text{or } D_{ij}^t = 0.00263 \pm 12\%.$$

The somewhat disappointing result of Example Problem 7.3 immediately leads us to make several important observations. Firstly, the result was obtained without any consideration of the details of the enclosure, its geometry, or its surface models other than the value of the distribution factor itself. Secondly, for the investment of 100 000 energy bundles one might reasonably expect to get a narrower 95% confidence interval than ±12%. Finally, it is clear from the form of Eq. (7.20) that interplay occurs between the value of the distribution factor estimate, $p = D^e_{ij}$, and the number of energy bundles emitted. To see this clearly, let us rewrite Eq. (7.20) in terms of the fractional width of the confidence interval,

$$1 - W_c\sqrt{\frac{1 - D^e_{ij}}{ND^e_{ij}}} \le \frac{D^t_{ij}}{D^e_{ij}} \le 1 + W_c\sqrt{\frac{1 - D^e_{ij}}{ND^e_{ij}}}, \tag{7.21}$$

where now D^i_{ij} is the (unknown) true value of the distribution factor and D^e_{ij} is the MCRT estimate based on tracing n energy bundles. It is clear from this form of the confidence interval expression that to minimize the departure of D^i_{ij}/D^e_{ij} from unity we must minimize the quantity

$$W_c\sqrt{\frac{1 - D^e_{ij}}{ND^e_{ij}}}. \tag{7.22}$$

For sufficiently large sample sizes N the critical value of the W statistic is fixed once we state our desired level of confidence. Therefore, all that remains to work with is the quantity

$$f(N, D^e_{ij}) = \sqrt{\frac{1 - D^e_{ij}}{ND^e_{ij}}}, \tag{7.23}$$

whose behavior is shown in Figure 7.4.

Figure 7.4 may be interpreted as showing the fractional uncertainty in the distribution factor per critical value of the W statistic for a given number of energy bundles traced. It is clear from the figure and from inspection of Eq. (7.23) that for very small values of the distribution factor the uncertainty can be very large. This is because the absorbing surface, surface j, is increasingly **undersampled** by the Monte Carlo ray-trace method as the value of the distribution factor decreases. Figure 7.4 also verifies the expectation that increasing the number of energy bundles traced reduces the uncertainty in the value of the distribution factor. While this result is not surprising, it is significant to note that the improvement achieved by increasing N is much greater for small values of the distribution factor than for large values.

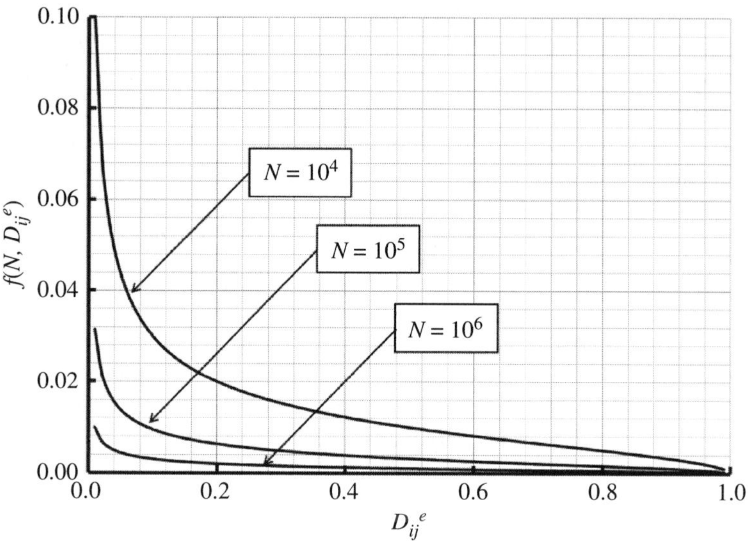

Figure 7.4 Behavior of $f(N, D^e_{ij})$ defined by Eq. (7.23).

7.5 Effects of Uncertainties in the Enclosure Geometry and Surface Models

In the previous section it was tacitly assumed that the estimates for the distribution factors were based on exactly known enclosure geometry and surface models. We refer to confidence intervals obtained under that assumption as **inherent confidence intervals**. That is, they are inherent to the statistical nature of the MCRT method and do not take into account our degree of confidence in the model upon which the Monte Carlo ray-trace is based. An imperfect model will lead to imperfect results for the distribution factors whose degree of imperfection goes beyond that implied by the width of the inherent confidence interval.

Now, in practice, enclosure dimensions and surface models carry their own uncertainties. Recall that the uncertainty in a quantity is the (unknown) difference between the true and observed values. At this point it is convenient to distinguish between the **inherent uncertainty** in the distribution factor estimate related to the inherent confidence interval developed in the previous section, and the **aggravated uncertainty** related to the uncertainties in the enclosure geometry and surface models. All uncertainties are themselves random variables and so are subject to statistical treatment. We seek a formalism that would allow the inherent uncertainty to be combined with the aggravated uncertainty in a statistically meaningful way.

7.6 Single-Sample Versus Multiple-Sample Experiments

Section 7.4 deals with what are called **single-sample experiments** because the statistical analysis leading to the uncertainty interval is based on only a single sample of size N. Now consider a situation in which M samples of a population are obtained, where each sample consists of N observations. Such experiments are called **multiple-sample experiments**. For a single-sample experiment the population mean μ and standard deviation σ are related to the **mean of the sample means** μ_m and the **standard deviation of the sample means** σ_m according to

$$\mu_m = \mu \tag{7.24}$$

and

$$\sigma_m = \frac{\sigma}{\sqrt{N}}. \tag{7.25}$$

These same quantities can be estimated from a multiple-sample experiment in which M independent sample means are obtained as

$$m_m = \frac{1}{M}\sum_{i=1}^{M} m_i \tag{7.26}$$

and

$$s_m = \sqrt{\frac{1}{M-1}\sum_{i=1}^{M}(m_i - m_m)^2}. \tag{7.27}$$

Similarly, the **mean of the sample proportions** μ_p and the **standard deviation of the sample proportions** σ_p are given by

$$\mu_p = \pi \tag{7.28}$$

and

$$\sigma_p = \sqrt{\frac{\pi(1-\pi)}{N}}. \tag{7.29}$$

These latter two quantities can be estimated from a multiple-sample experiment, in which M independent sample proportions are obtained, as

$$m_p = \frac{1}{M} \sum_{i=1}^{M} p_i \tag{7.30}$$

and

$$s_p = \sqrt{\frac{m_p(1 - m_p)}{MN}}. \tag{7.31}$$

Note that in the special case where $M = 1$ (i.e., a single-sample experiment), Eq. (7.31) reduces to Eq. (7.18).

Now recall that an estimate of σ_p was used in the single-sample experiment as the basis for establishing the inherent confidence interval in Section 7.4. In the case of a multiple-sample experiment we use Eq. (7.30) as the estimate of μ_p, and Eq. (7.31) as the estimate of σ_p to establish the corresponding confidence interval,

$$m_p - W_c s_p \leq \pi \leq m_p + W_c s_p. \tag{7.32}$$

As a practical matter an M-sample experiment in the MCRT sense is completely equivalent to a single-sample experiment with $N' = MN$. Therefore, it should not be surprising that the confidence interval obtained using Eq. (7.20) with N' observations is the same as that obtained using Eq. (7.32) using M samples, each consisting of N observations, provided that the same experiment is sampled each time. Then what is gained by performing a multiple-sample experiment (other than the obvious fact that the uncertainty interval is smaller)? In the next section we learn that multiple-sample experiments provide a convenient means for evaluating the aggravated uncertainty of the distribution factor estimate.

7.7 Evaluation of Aggravated Uncertainty

In the previous section it was established that for a multiple-sample experiment having M samples, each consisting of N observations, the uncertainty interval expression is equivalent to that for a single-sample experiment consisting of $N' = MN$ observations. However, it was tacitly assumed in the development that the aggravated uncertainty was zero; that is, that the same experiment was sampled each time. But suppose this is not the case? In fact, Eqs. (7.30)–(7.32) still apply except that now the variation in the estimator for the population proportion p from experiment to experiment can include the effects of the uncertainties in parameters describing the experiment.

In the context of a Monte Carlo ray-trace, it must be recognized that the enclosure dimensions and surface optical models themselves are susceptible to random variations. The approach in this case is to run a series of M numerical experiments for which the dimensions and the coefficients characterizing the surface optical models vary normally, with specified means and standard deviations, from experiment to experiment. As an example, consider the directional, band-averaged spectral absorptivity. Two successive normally distributed perturbed values of $\alpha_{ik}(\vartheta, \varphi)$ may be related to the mean value $\mu_{ik}(\vartheta, \varphi)$ and the standard deviation $\sigma_{ik}(\vartheta, \varphi)$ according to [2]

$$\alpha_{ik}(\vartheta, \varphi) = \mu_{ik}(\vartheta, \varphi) + \sigma_{ik}(\vartheta, \varphi)\sqrt{2\ln(1/R_1)}\cos(2\pi R_2) \qquad (7.33a)$$

and

$$\alpha_{ik}(\vartheta, \varphi) = \mu_{ik}(\vartheta, \varphi) + \sigma_{ik}(\vartheta, \varphi)\sqrt{2\ln(1/R_1)}\sin(2\pi R_2), \qquad (7.33b)$$

where R_1 and R_2 are two random numbers whose values are uniformly distributed between zero and unity. Equation (7.33) is used to pre-assign the emissivity, absorptivity, and reflectivity of all n surfaces before each of the M experiments. A similar treatment can be applied to key dimensions of the enclosure. Then, after a statistically significant number M of numerical experiments have been carried out, Eqs. (7.30)–(7.32) can be used to estimate the aggravated uncertainty of the distribution factors.

This approach for estimating the aggravated uncertainty of the distribution factors is potentially very costly. Firstly, we must enssure that the number of energy bundles emitted in a given experiment is statistically significant. To accomplish this we follow the procedure described in Section 7.4 for single-sample experiments. Then we must repeat the experiment a sufficiently large number of times M to obtain a statistically meaningful distribution of population proportion estimates p. There now remains the question, "How are the uncertainties in the distribution factor estimates used to determine the uncertainties in the temperatures and the heat transfer rates obtained using the distribution factors?"

7.8 Uncertainty in Temperature and Heat Transfer Results

The celebrated 1953 article by Kline and McClintock [3] provides a widely accepted formalism for relating the uncertainty of an experimental result, ω_R, to the uncertainties of the independent component measurements used to calculate the result. The procedure, which is equally applicable to physical experiments and numerical experiments of the type considered in this book, may be summarized as follows.

Suppose a result R is related to a series of n measurement variables X_i, $i = 1, 2, \ldots, n$ according to

$$R = R(X_1, X_2, \ldots, X_n). \tag{7.34}$$

Then, following the formalism of Kline and McClintock,

$$\omega_R = \pm\sqrt{\left(\frac{\partial R}{\partial X_1}\omega_{X_1}\right)^2 + \left(\frac{\partial R}{\partial X_2}\omega_{X_2}\right)^2 + \cdots + \left(\frac{\partial R}{\partial X_n}\omega_{X_n}\right)^2}, \tag{7.35}$$

in which the ω_i are the uncertainties of the component measurement variables X_i. Three observations are appropriate at this point. Firstly, for Eq. (7.34) to be valid, all of the component uncertainties must be stated with the same level of confidence. Implied is that they are each random variables and so are normally distributed. Secondly, the existence of an explicit analytical expression relating the result to the component measurements, that is Eq. (7.34), is implied. This requirement is needed to allow the partial derivatives ("sensitivities") in Eq. (7.35) to be evaluated. For problems amenable to solution by the MCRT method, the explicit analytical expressions depend on the complexity of the problem being solved. Thirdly, it is tacitly assumed that the various measurement variables are independent of each other. In the following discussion, application of the method of Kline and McClintock is illustrated for the case of a total, diffuse-specular analysis as developed and illustrated in Chapter 3. However, the extension of the method to the material in Chapters 4–6 is straightforward.

We consider the case where a diffuse-specular gray enclosure has specified net heat fluxes on surface elements $1, 2, \ldots, N$ and specified temperatures on surface elements $N+1, N+2, \ldots, n$. In this situation Eq. (3.38) for the unknown surface temperatures can be written

$$T_i = \sqrt[4]{\frac{1}{\sigma}\sum_{j=1}^{N}\left\{[\Psi_{ij}]^{-1}\left[q_j - \varepsilon_j \sum_{i=N+1}^{n} \sigma T_j^4(\delta_{ji} - D_{ji})\right]\right\}}, \quad 1 \le i \le N, \tag{7.36}$$

where

$$\Psi_{ij} = \varepsilon_i(\delta_{ij} - D_{ij}), \quad 1 \le i \le N, \quad 1 \le j \le N, \tag{3.41}$$

and Eq. (3.35) for the unknown surface heat fluxes can be written

$$q_i = \varepsilon_i \sum_{j=1}^{n} \sigma T_j^4(\delta_{ij} - D_{ij}), \quad N + 1 \leq i \leq n. \tag{7.37}$$

In Eqs. (7.36), (3.41), and (7.37), and in the rest of this chapter, it is understood that D_{ij} is the *estimate* of the total, diffuse-specular radiation distribution factor.

It is assumed in the following development that each of the "known" surface temperatures T_i has associated with it an uncertainty ω_T and that each "known" surface net heat flux q_i has associated with it an uncertainty ω_q. This is in addition to the uncertainties ω_ε associated with the emissivities ε and the uncertainties ω_D associated with the distribution factors D_{ij}. It must be noted here that the distribution factors are, in fact, not independent of the emissivities, even though this is a formal requirement when treating both as measurement variables in the Kline and McClintock formulation. This requirement may be ignored here because, while q_i depends directly and explicitly on ε_i, the dependence of D_{ij} on ε_i is indirect (through reflections), is distributed across all the surfaces, and is statistically weak. Finally, it is assumed that the confidence interval is the same, usually 95%, on all of these uncertainties.

Because of its form, Eq. (7.36) presents a particularly nasty challenge in determining the sensitivities $\partial T_i/\partial T_{i>N}$ and $\partial T_i/\partial q_j$ and especially $\partial T_i/\partial D_{ij}$ and $\partial T_i/\partial \varepsilon_j$. Evaluation of the sensitivities $\partial q_i/D_{ij}$, $\partial q_i/\partial \varepsilon_i$, and $\partial q_i/\partial T_j$ on the other hand is straightforward. When analytical expressions for these sensitivities have been obtained, their values are computed for each surface or volume element for which unknown surface temperatures and surface net heat fluxes are to be found.

7.9 Application to the Case of Specified Surface Temperatures

The following analysis closely follows Chapter 3 in M.C. Sanchez's doctoral dissertation [4], which includes several corrections to and improvements over the treatment in Ref. [5]. We consider the case in which all of the surface temperatures are specified, with appropriate uncertainties. In this situation Eq. (7.37) can be written

$$q_i = q_{i,e} - q_{i,a}, \quad 1 \leq i \leq n, \tag{7.38}$$

where

$$q_{i,e} = \varepsilon_i \sigma T_i^4 \tag{7.39}$$

and

$$q_{i,a} = \varepsilon_i \sum_{j=1}^{n} \sigma T_j^4 D_{ij}. \tag{7.40}$$

Equation (7.35) can then be written

$$\omega_{q_i} = \pm\sqrt{\left(\frac{\partial q_i}{\partial q_{i,e}}\omega_{q_{i,e}}\right)^2 + \left(\frac{\partial q_i}{\partial q_{i,a}}\omega_{q_{i,a}}\right)^2}, \quad 1 \le i \le n, \tag{7.41}$$

where

$$\omega_{q_{i,e}} = \pm\sqrt{\left(\frac{\partial q_{i,e}}{\partial \varepsilon_i}\omega_{\varepsilon_i}\right)^2 + \left(\frac{\partial q_{i,e}}{\partial T_i}\omega_{T_i}\right)^2} \tag{7.42}$$

and

$$\omega_{q_{i,a}} = \pm\sqrt{\left(\frac{\partial q_{i,a}}{\partial \varepsilon_i}\omega_{\varepsilon_i}\right)^2 + \left(\frac{\partial q_{i,a}}{\partial T_j}\omega_{T_j}\right)^2 + \left(\frac{\partial q_{i,a}}{\partial D_{ij}}\omega_{D_{ij}}\right)^2}. \tag{7.43}$$

The sensitivities in Eqs. (7.41)–(7.43) are

$$\frac{\partial q_i}{\partial q_{i,e}} = 1, \quad \frac{\partial q_i}{\partial q_{i,a}} = -1, \quad \frac{\partial q_{i,e}}{\partial \varepsilon_i} = \sigma T_i^4, \quad \frac{\partial q_{i,e}}{\partial T_i} = 4\varepsilon_i \sigma T_i^3,$$

$$\frac{\partial q_{i,a}}{\partial \varepsilon_i} = \sum_{j=1}^{n} \sigma T_j^4 D_{ij}, \quad \frac{\partial q_{i,a}}{\partial T_j} = 4\varepsilon_i \sum_{j=1}^{n} \sigma T_j^3 D_{ij}, \tag{7.44}$$

and

$$\frac{\partial q_{i,a}}{\partial D_{ij}} = \varepsilon_i \sum_{j=1}^{n} \sigma T_j^4.$$

With the introduction of Eqs. (7.42)–(7.44), Eq. (7.41) may be written

$$\omega_{q_i} = \pm\left\{(\sigma T_i^4 \omega_{\varepsilon_i})^2 + (4\varepsilon_i \sigma T_i^3 \omega_{T_i})^2 + \left(\omega_{\varepsilon_i} \sum_{j=1}^{n} \sigma T_j^4 D_{ij}\right)^2 \right.$$

$$\left. + \left(4\varepsilon_i \sum_{j=1}^{n} \sigma T_j^3 \omega_{T_j} D_{ij}\right)^2 + \left(\varepsilon_i \sum_{j=1}^{n} \sigma T_j^4 \omega_{D_{ij}}\right)^2 \right\}^{\frac{1}{2}}, \quad 1 \le i \le n. \tag{7.45}$$

Finally, by convention we define the *relative* uncertainty in the net heat flux from surface element i as a fraction of the heat flux *emitted from i*; that is,

$$\frac{\omega_{q_i}}{\varepsilon_i \sigma T_i^4} = \pm \left\{ \left(\frac{\omega_{\varepsilon_i}}{\varepsilon_i}\right)^2 + \left(4\frac{\omega_{T_i}}{T_i}\right)^2 + \left[\frac{\omega_{\varepsilon_i}}{\varepsilon_i}\sum_{j=1}^{n}\left(\frac{T_j}{T_i}\right)^4 D_{ij}\right]^2 \right.$$
$$\left. + \left[4\sum_{j=1}^{n}\left(\frac{T_j}{T_i}\right)^4 \frac{\omega_{T_j}}{T_j} D_{ij}\right]^2 + \left[\sum_{j=1}^{n}\left(\frac{T_j}{T_i}\right)^4 \omega_{D_{ij}}\right]^2 \right\}^{1/2},$$
$$1 \le i \le n. \quad (7.46)$$

The result represented by Eq. (7.46) is exact; that is, no simplifying assumptions have been invoked to this point. It is reasonable to assume that the fractional (percentage) uncertainties of temperature, emissivity, and distribution factor are the same for all surface elements of the enclosure; i.e., $\omega_{T_j}/T_i = \omega_T/T$, $\omega_{\varepsilon_i}/\varepsilon_i = \omega_\varepsilon/\varepsilon$, and $\omega_{D_{ij}}/D_{ij} = \omega_D/D$, where each is a generally different constant. In this case Eq. (7.46) becomes

$$\frac{\omega_{q_i}}{\varepsilon_i \sigma T_i^4} = \pm \left\{ \left[\left(\frac{\omega_\varepsilon}{\varepsilon}\right)^2 + \left(4\frac{\omega_T}{T}\right)^2\right]\left[1 + \left(\sum_{j=1}^{n}\left(\frac{T_j}{T_i}\right)^4 D_{ij}\right)^2\right] \right.$$
$$\left. + \left(\frac{\omega_D}{D}\right)^2 \left[\sum_{j=1}^{n}\left(\frac{T_j}{T_i}\right)^4 D_{ij}\right]^2 \right\}^{1/2}, \quad 1 \le i \le n. \quad (7.47)$$

Equation (7.47) illustrates the important role played by the **temperature spread** T_j/T_i in the uncertainty of the relative net heat flux. Consideration of the special case of an isothermal enclosure ($T_j/T_i = 1$ for all i and j) leads to

$$\frac{\omega_{q_i}}{\varepsilon_i \sigma T_i^4} = \pm\sqrt{2\left[\left(\frac{\omega_\varepsilon}{\varepsilon}\right)^2 + \left(4\frac{\omega_T}{T}\right)^2\right] + \left(\frac{\omega_D}{D}\right)^2},$$
$$1 \le i \le n \text{ (isothermal enclosure).} \quad (7.48)$$

Sanchez [4] has demonstrated through numerous numerical experiments that the uncertainty estimate obtained using Eq. (7.47) is minimized for the case of an isothermal enclosure. We conclude that, as the temperature distribution in an enclosure becomes more uniform, the validity of the approximation represented by Eq. (7.48) increases.

In summary, Eq. (7.46) can be used to estimate, to a stated level of confidence, the relative uncertainty in the net heat flux from surface element i if all of the surface temperatures, emissivities, and surface-to-surface distribution factors are known to the same level of confidence. The somewhat simpler Eq. (7.47) can be used when the relative uncertainties of the surface temperatures, emissivities, and the distribution factors are the same in each category, and are known to the same degree of confidence. Treatment of the case where surface net heat flux is specified for one or more surfaces, as developed in Section 3.7, is left as an exercise.

7.10 Experimental Design of MCRT Algorithms

We can now apply the statistical principles developed in this chapter to the design of MCRT algorithms. Here the term "design" is employed in the restricted sense used by statisticians; that is, we seek the number of surface elements n and the number of energy bundles N traced per surface element required to attain, to a stated level of confidence, a specified uncertainty in the surface net heat flux results.

We continue to consider the special case of a diffuse-specular gray enclosure for which the temperature is specified on all of the surface elements, and we proceed under the assumption that the relative uncertainties in surface temperature and surface emissivity, and the desired uncertainty in surface net heat flux relative to the local emitted flux, have been specified in advance, all to the same level of confidence. In the present development we assume further that the relative uncertainties are the same within a given category; that is, the development is based on Eq. (7.47).

Experimental design in the present context may be divided into three steps: (i) determine the number of surface elements n into which the enclosure must be divided, (ii) determine the relative uncertainty in the distribution factors consistent with obtaining the desired uncertainty in the surface net heat fluxes, and (iii) determine the number of energy bundles that must be traced per surface element to achieve the required uncertainty in the distribution factors.

***Step 1*: Determine the number n of surface elements to be used.**

The number of surface elements n to be used is usually dictated by a trade-off between the desired spatial resolution of the result on the one hand and the available computer resources on the other. Execution time is roughly proportional to n^2, while spatial resolution is a linear function of n. Therefore, the cost associated with increasing n increases much more rapidly than the benefit. The final decision should also be informed by the spatial resolution of the given temperature and/or net heat flux distributions. It is pointless to seek higher resolution in the unknown temperature and net heat flux fields than is available in the known distributions.

***Step 2*: Determine the mean value of the relative uncertainty in the distribution factors, ω_D/D.**

In this step we seek the value of ω_D/D that produces the desired *mean* value of $\omega_{q_i}/\varepsilon_i\sigma T_i^4$,

$$\langle \omega_q/\varepsilon\sigma T^4 \rangle \equiv \frac{1}{n}\sum_{i=1}^{n}\frac{\omega_{q_i}}{\varepsilon_i\sigma T_i^4}. \tag{7.49}$$

Use of Eq. (7.49) with Eq. (7.47) to determine $\langle \omega_q/\varepsilon\sigma T^4 \rangle$ for each candidate value of ω_D/D would require knowledge of the distribution factors, which implies prior knowledge of the number N of rays traced. But determination of N is precisely one of the goals of experimental design. Therefore, we seek an alternative means to estimate, or at least to bound, the value of $\langle \omega_q/\varepsilon\sigma T^4 \rangle$ without knowledge of the distribution factors themselves.

We begin by squaring both sides of Eq. (7.47),

$$\left(\frac{\omega_{q_i}}{\varepsilon_i\sigma T_i^4}\right)^2 = \left[\left(\frac{\omega_\varepsilon}{\varepsilon}\right)^2 + \left(4\frac{\omega_T}{T}\right)^2\right]\left[1 + \left(\sum_{j=1}^{n}\left(\frac{T_j}{T_i}\right)^4 D_{ij}\right)^2\right] + \left(\frac{\omega_D}{D}\right)^2\left[\sum_{j=1}^{n}\left(\frac{T_j}{T_i}\right)^4 D_{ij}\right]^2 \quad 1 \le i \le n. \tag{7.50}$$

We then replace D_{ij} with its mean value, $1/n$. This allows us to replace $\sum_{j=1}^{n}(T_j/T_i)^4 D_{ij}$ with the mean value over j of $(T_j/T_i)^4$, $\langle (T_j/T_i)^4 \rangle_j$. Thus, to the extent that replacement of D_{ij} by its mean value is valid, we have the approximation

$$\left(\frac{\omega_{q_i}}{\varepsilon_i\sigma T_i^4}\right)^2 \approx \left[\left(\frac{\omega_\varepsilon}{\varepsilon}\right)^2 + \left(4\frac{\omega_T}{T}\right)^2\right]\left[1+\left(\left\langle\left(\frac{T_j}{T_i}\right)^4\right\rangle_j\right)^2\right] + \left(\frac{\omega_D}{D}\right)^2\left(\left\langle\left(\frac{T_j}{T_i}\right)^4\right\rangle_j\right)^2. \tag{7.51}$$

If we now replace $(\omega_{q_i}/\varepsilon_i\sigma T_i^4)$ and $\langle(T_j/T_i)^4\rangle_j$ by their mean values over i we obtain the further approximation

$$\left\langle\frac{\omega_{q_i}}{\varepsilon_i\sigma T_i^4}\right\rangle_i \approx \pm\sqrt{\begin{matrix}\left[\left(\frac{\omega_\varepsilon}{\varepsilon}\right)^2 + \left(4\frac{\omega_T}{T}\right)^2\right]\left[1+\left(\left\langle\left\langle\left(\frac{T_j}{T_i}\right)^4\right\rangle\right\rangle\right)^2\right] \\ +\left(\frac{\omega_D}{D}\right)^2\left(\left\langle\left\langle\left(\frac{T_j}{T_i}\right)^4\right\rangle\right\rangle\right)^2\end{matrix}}, \tag{7.52}$$

where

$$\left\langle\left\langle\left(\frac{T_j}{T_i}\right)^4\right\rangle\right\rangle = \left\langle\left\langle\left(\frac{T_j}{T_i}\right)^4\right\rangle_j\right\rangle_i = \frac{1}{n^2}\sum_{i=1}^{n}\sum_{j=1}^{n}(T_j/T_i)^4. \tag{7.53}$$

The intended use of Eq. (7.52) is to facilitate estimation of the relative uncertainty in the distribution factors, ω_D/D, corresponding to the specified value of relative uncertainty in the heat transfer results. As a reality check we might note that Eq. (7.52) converges to Eq. (7.48) in the limit as $\langle\langle(T_j/T_i)^4\rangle\rangle$ approaches unity. After evaluating the actual mean fourth-power temperature spread from the known value of n and the specified temperature distribution, we find the value of ω_D/D which satisfies Eq. (7.52) consistent with the specified value of $\langle\omega_{q_i}/\varepsilon_i\sigma T_i^4\rangle_i$ (see Problem 7.9).

Figure 7.5 illustrates the dependence of the mean relative uncertainty of the net heat flux in an enclosure, Eq. (7.52), on the mean fourth-power temperature spread, Eq. (7.53), for a range of values of the mean relative uncertainty in the distribution factors, and for $\omega_\varepsilon/\varepsilon$ and ω_T/T both equal to 0.01. Note that this figure can be created without specific knowledge of the enclosure geometry or number of surface elements. Inspection of the figure reveals that the mean relative uncertainty in the net heat flux

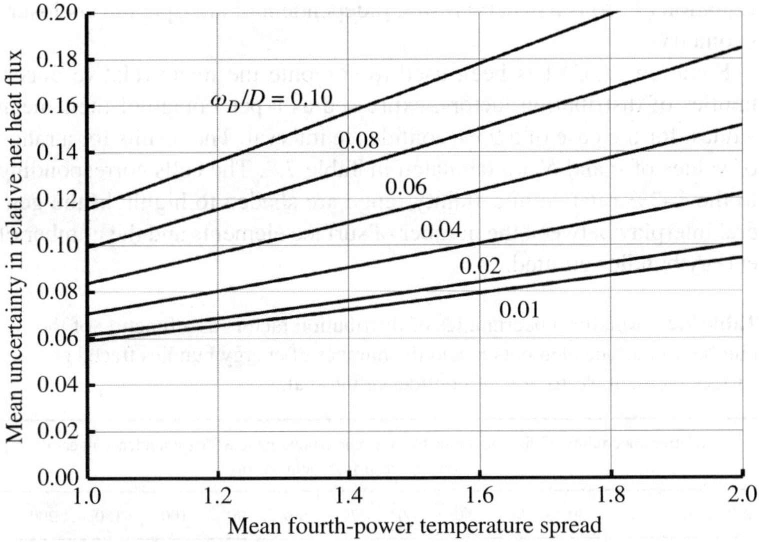

Figure 7.5 Variation of the mean relative uncertainty in net heat flux in an enclosure with mean fourth-power temperature spread for a range of values of ω_D/D ($\omega_\varepsilon/\varepsilon = \omega_T/T = 0.01$).

can be no less than about 6% even if the uncertainties in the emissivity, temperature, and distribution factors are limited to 1%.

***Step 3*: Determine the number of rays that must be traced per surface element.**

Referring to Eq. (7.21), we recognize that the uncertainty in the distribution factors is related to their estimated values and to the related confidence interval according to

$$\frac{\omega_{D_{ij}}}{D_{ij}} = \pm W_c \sqrt{\frac{(1 - D_{ij})}{ND_{ij}}}. \tag{7.54}$$

Equation (7.54) is valid for a particular distribution factor between two specified surface elements i and j, but we need an expression for the mean value of relative uncertainty, ω_D/D. With the introduction of the mean value of all distribution factors, $\langle D_{ij} \rangle = 1/n$, Eq. (7.54) becomes

$$\left\langle \left\langle \frac{\omega_{D_{ij}}}{D_{ij}} \right\rangle \right\rangle \equiv \frac{\omega_D}{D} \approx \pm W_c \sqrt{\frac{n-1}{N}}. \tag{7.55}$$

Equation (7.55) is a general result, independent of any specific enclosure geometry.

Equation (7.55) has been used to compute the mean relative uncertainties of distribution factors, expressed as a percentage of their mean values, for the case of a 95% confidence interval. The results for a range of values of n and N are tabulated in Table 7.2. The cells corresponding to the 1–2% relative uncertainty range are shaded to highlight the general interplay between the number of surface elements and the number of energy bundles emitted.

Table 7.2 Relative uncertainties of distribution factors as a function of the number of surface elements n, and the number of energy bundles traced per surface element, N, for a 95% confidence interval.

Relative uncertainty of distribution factors (%) corresponding to a 95% confidence interval												
	Number of surface elements (n)											
$N(10^3)$	10	20	30	40	50	60	70	80	90	100	150	200
10	5.880	8.543	10.555	12.240	13.720	15.055	16.281	17.421	18.491	19.502	23.925	27.649
20	4.158	6.041	7.463	8.655	9.702	10.646	11.512	12.318	13.075	13.790	16.917	19.551
50	2.630	3.821	4.720	5.474	6.136	6.733	7.281	7.791	8.269	8.721	10.700	12.365
100	1.859	2.702	3.338	3.871	4.339	4.761	5.148	5.509	5.847	6.167	7.566	8.743
150	1.518	2.206	2.725	3.160	3.542	3.887	4.204	4.498	4.774	5.035	6.177	7.139
200	1.315	1.910	2.360	2.737	3.068	3.366	3.641	3.895	4.135	4.361	5.350	6.183
250	1.176	1.709	2.111	2.448	2.744	3.011	3.256	3.484	3.698	3.900	4.785	5.530
300	1.074	1.560	1.927	2.235	2.505	2.749	2.972	3.181	3.376	3.561	4.368	5.048
350	0.994	1.444	1.784	2.069	2.319	2.545	2.752	2.945	3.125	3.296	4.044	4.674
400	0.930	1.351	1.669	1.935	2.169	2.380	2.574	2.754	2.924	3.083	3.783	4.372
450	0.877	1.274	1.573	1.825	2.045	2.244	2.427	2.597	2.756	2.907	3.567	4.122
500	0.832	1.208	1.493	1.731	1.940	2.129	2.302	2.464	2.615	2.758	3.383	3.910
550	0.793	1.152	1.423	1.650	1.850	2.030	2.195	2.349	2.493	2.630	3.226	3.728
600	0.759	1.103	1.363	1.580	1.771	1.944	2.102	2.249	2.387	2.518	3.089	3.569
650	0.729	1.060	1.309	1.518	1.702	1.867	2.019	2.161	2.293	2.419	2.968	3.429
700	0.703	1.021	1.262	1.463	1.640	1.799	1.946	2.082	2.210	2.331	2.860	3.305
750	0.679	0.987	1.219	1.413	1.584	1.738	1.880	2.012	2.135	2.252	2.763	3.193
800	0.657	0.955	1.180	1.368	1.534	1.683	1.820	1.948	2.067	2.180	2.675	3.091
850	0.638	0.927	1.145	1.328	1.488	1.633	1.766	1.890	2.006	2.115	2.595	2.999
900	0.620	0.901	1.113	1.290	1.446	1.587	1.716	1.836	1.949	2.056	2.522	2.914
950	0.603	0.877	1.083	1.256	1.408	1.545	1.670	1.787	1.897	2.001	2.455	2.837
1000	0.588	0.854	1.055	1.224	1.372	1.506	1.628	1.742	1.849	1.950	2.392	2.765
1500	0.480	0.698	0.862	0.999	1.120	1.229	1.329	1.422	1.510	1.592	1.953	2.258
2000	0.416	0.604	0.746	0.866	0.970	1.065	1.151	1.232	1.307	1.379	1.692	1.955
2500	0.372	0.540	0.668	0.774	0.868	0.952	1.030	1.102	1.169	1.233	1.513	1.749
3000	0.339	0.493	0.609	0.707	0.792	0.869	0.940	1.006	1.068	1.126	1.381	1.596
3500	0.314	0.457	0.564	0.654	0.733	0.805	0.870	0.931	0.988	1.042	1.279	1.478
4000	0.294	0.427	0.528	0.612	0.686	0.753	0.814	0.871	0.925	0.975	1.196	1.382
4500	0.277	0.403	0.498	0.577	0.647	0.710	0.767	0.821	0.872	0.919	1.128	1.303
5000	0.263	0.382	0.472	0.547	0.614	0.673	0.728	0.779	0.827	0.872	1.070	1.237

Example Problem 7.4

An MCRT model is to be used to estimate the surface net heat fluxes for an enclosure consisting of 100 surface elements, all of whose temperatures are known. How many energy bundles must be traced per surface element if the upper limit on the mean value of the relative uncertainty of the distribution factors needs to be 0.01 (1%) to achieve the desired limit on the mean uncertainty in the surface net heat fluxes? Assume that a 95% level of confidence is required.

Solution

Applying Eq. (7.55), we obtain

$$N = \frac{W_c^2(n-1)}{(\omega_D/D)^2}\,\frac{(1.960)^2(100-1)}{(0.01)^2} \approx 3.8 \times 10^6. \qquad (7.56)$$

We would have to trace about four million energy bundles from each surface element to assure that, to a 95% level of confidence, the distribution factors were, on the average, obtained to within a 1% uncertainty.

Example Problem 7.5

Consider the cubic enclosure, illustrated in Figure 1.16, which is the subject of Problems 1.25, 1.26, 3.15–3.19, and 4.11–4.14. Suppose that one-half of the surfaces have a temperature of 500 K and the other half have a temperature of 600 K. (i) Compute the mean fourth-power temperature spread in this enclosure. (ii) Estimate the mean relative uncertainty in net heat flux from the enclosure surfaces if the mean relative uncertainties in emissivities, temperatures, and distribution factors are all 1 %. (iii) What is the minimum number of rays that must be be traced from each surface to attain this value of mean relative uncertainty in net heat flux?

Solution

Using a spreadsheet, the mean fourth-power temperature spread is $\langle\langle(T_j/T_i)^4\rangle\rangle = 1.139$; the mean relative uncertainty in net heat flux is 0.0635, or 6.35%; and 883 568 rays must be traced from each surface element to obtain a 1% mean relative uncertainty in the distribution factors.

Problems

7.1 Rework Example Problem 7.1 using a sequence of 100 pseudo-random numbers from Excel or a similar source. What may you conclude about the pseudo-random number generator?

7.2 Assuming that when you worked Problem 7.1 you obtained a result suggesting that your pseudo-random number generator may be biased toward either zero or unity, perform the appropriate hypothesis test to either verify or dispel this suggestion. [*Hint*: See Example Problem 7.2.]

7.3 Draw ten 100-member pseudo-random number sequences from Excel or a similar source. Then create a table similar to Table 7.1. What is the mean of the means of the 10 sequences? What may you conclude from this experiment about your pseudo-random number generator?

7.4 Use the binomial probability distribution to compute the probability of rolling "boxcars" (two sixes) using a pair of fair dice.

7.5 Derive an expression for the sensitivity $\partial T_i / \partial q_j$, $1 \le i \le N$, where the dependence of T_i on q_j is as given by Eq. (7.36).

7.6 Show formally that the mean value of all of the distribution factors in an enclosure (the "global" mean) is $1/n$, where n is the number of surface elements defining the enclosure.

7.7 Prove that the minimum possible value of the mean fourth-power temperature spread, defined by Eq. (7.53), is unity.

7.8 Randomly assign temperatures ranging between 1000 and 100 K to an enclosure consisting of 10 surface elements. Then compute the value of the mean fourth-power temperature spread defined by Eq. (7.53). Then repeat these calculations with the temperatures ranging randomly between 500 and 600 K. What may we conclude from these calculations?

7.9 Carefully solve Eq. (7.52) algebraically for ω_D/D. Note that all the terms under the radical in the result must carry a positive (+) sign. How do we justify this?

7.10 Based on the results of Problem 7.8, use Eq. (7.52) to estimate the value of the relative uncertainty in the distribution factors, ω_D/D, corresponding to a specified value of relative uncertainty in the heat transfer results of 6%. Assume the mean uncertainties in emissivity and temperature are both 1%.

7.11 A MCRT model is to be used to estimate the surface net heat fluxes for an enclosure consisting of 24 surface elements, all of whose temperatures are known. How many energy bundles must be traced per surface element if the upper limit on the mean value of the relative uncertainty of the distribution factors needs to be 0.02 (2%) to achieve the specified mean uncertainty in the surface net heat fluxes? Assume that a 95% level of confidence is required.

7.12 For the enclosure in Example Problem 7.5, let the surfaces lying in the planes containing the point (0, 0, 0) have a mean temperature of 500 K with a standard deviation of 5 K, while the surfaces lying in the planes containing the point (Δx, Δy, Δz) have a mean temperature of 600 K, also with a standard deviation of 5 K. Note that Eq. (7.33) can be used to compute two successive values of temperature. For the purposes of this exercise we will identify the uncertainty in the local surface temperature with its standard deviation ($k = 1$).

(a) Perform a Monte Carlo ray-trace analysis in which you trace one million rays from each surface element to determine the net heat flux from each surface element. Assume that all surfaces are diffuse gray with a mean emissivity of 0.55 and a standard deviation of 0.01. Once again, use Eq. (7.33) to compute the local emissivity for each ray and the local absorptivity for each incidence of a ray to a surface.

(b) Use Eq. (7.47) to determine the relative uncertainty in the results obtained in Part (a).

(c) Compute the mean of the results obtained in Part (b). How does this result compare with the value predicted using Eq. (7.52)?

References

1. Koopmans, L.H. (1987). *Introduction to Contemporary Statistical Methods*, 2e, 206–207. Boston, MA: Duxbury.
2. Press, W.H., Teukolsky, S.A., Vetterling, W.T., and Flannery, B.P. (1992). *Numerical Recipes in C: The Art of Scientific Computing*, 2e, 288–290. New York: Cambridge University Press.
3. Kline, S.J. and McClintock, F.A. (January 1953). Describing uncertainties in single-sample experiments. *Mechanical Engineering* 1: 3.
4. Sanchez, M.C., (2002). Statistical analysis of the Monte Carlo ray-trace in radiation heat transfer. PhD dissertation, Virginia Tech.
5. Mahan, J.R. (2002). *Radiation Heat Transfer: A Statistical Approach*, New York: Wiley.

A
Appendix

Random Number Generators and Autoregression Analysis

The Monte Carlo ray-trace (MCRT) approach to radiation heat transfer analysis is highly dependent on the availability of a fast, efficient random number generator. In this appendix we learn that true random number generators do not exist, but rather that "random" numbers are computed using completely deterministic algorithms. In fact, the very success of the MCRT method requires that the same identical sequence of random numbers be available upon demand. The requirements for these pseudo-random number generators are established and an algorithm for computing them is identified. Finally, a brief introduction is provided to autoregression analysis, which is a useful tool for testing the randomness of pseudo-random number generators. Many readers will find autoregression analysis useful in other aspects of their work.

The Monte Carlo Ray-Trace Method in Radiation Heat Transfer and Applied Optics,
First Edition. J. Robert Mahan.

This Work is a co-publication between John Wiley & Sons Ltd and ASME Press.
Companion website: www.wiley.com/go/mahan/monte-carlo

A.1 Pseudo-Random Number Generators

The accuracy of a Monte Carlo simulation depends on the quality of the random number generator used. In the early days of computing, physical devices that supplied truly random numbers were attached to computers. Examples include devices for sensing Johnson ("shot") noise in an electronic circuit and for counting particles per unit time associated with radioactive decay. This approach was soon found to be impractical, mostly because of the inherently low rate at which these devices were capable of producing a sequence of random numbers.

Physical devices were eventually replaced with **pseudo-random** number generators (PRNGs), which use deterministic formulas to produce sequences of numbers whose statistical properties approach those of true random number sequences. Unfortunately, a correlation necessarily exists among members of a given sequence of numbers produced by a pseudo-random number generator; that is, knowledge of the value of a given member or members of the sequence is sufficient to predict the value of a later member in the sequence. Also, pseudo-random number generators are by their nature periodic; that is, the sequence they produce eventually repeats itself. If the periodicity of a given PRNG happens to match that of the sequence or sequences in which the random numbers are used in a simulation, biases may occur in the simulated result.

An alternative to the PRNG is the **quasi-random** number generator (QRNG), which produces sequences not intended to be random but rather intended to be distributed as uniformly as possible. In some Monte Carlo simulations, QRNGs may be more appropriate than PRNGs. However, in the radiation heat transfer calculations developed in this book PRNGs usually are the appropriate choice.

A.2 Properties of a "Good" Pseudo-Random Number Generator

A "good" pseudo-random number generator suitable for radiation heat transfer analysis will exhibit the following seven properties (adapted from James [1]).

A.2.1 Property 1

The sequence of numbers generated will be **uniformly distributed**. Uniformity of distribution of a sequence of pseudo-random numbers may be tested using the standard χ^2 test from statistics, which compares the observed and expected frequencies of the pseudo-random numbers in the sequence. In this test a sequence of pseudo-random numbers of length N is sorted according to the range into which its individual members fall. For example, suppose $N = 10\,000$ and a given member of the complete sequence has a value between 0.00 … and 0.05 … (the ellipses "…" indicate an unspecified number of zeros, i.e., an unspecified precision). Then that number is placed in bin 1, while any number between 0.05 … and 0.10 … is placed in bin 2, and so forth. When all 10 000 numbers have been sorted into the 20 bins, the count of numbers in each bin is obtained. Of course, the expected count for each bin is 10 000/20 = 500. We then compute

$$\chi^2 = \sum_{i=1}^{20} \left(\frac{O_i - E_i}{E_i} \right)^2, \tag{A.1}$$

where 0_i is the observed count in bin i and E_i is the expected count in bin i. The value of χ^2 may then be interpreted as a measure of whether or not the random numbers are uniformly distributed. The probability that the random number distribution is uniform is read from a chart, available in standard texts such as Ref. [2] (p. 541). To enter the chart we need the value of χ^2 and the number of **degrees of freedom** $d = n - k$, where n is the number of observations (20 in the current example) and k is the number of conditions imposed on the distribution (one in this case: uniformity). For $d = 19$, the value of χ^2 would have to be about 30 to have a 95% confidence that this distribution was uniform.

A.2.2 Property 2

The sequence of numbers generated will **"appear" random**; that is, it will pass any statistical test for randomness. The correlation between each number generated and the numbers that precede it in the sequence should be low. This means that knowledge of the value of any one member or subsequence of members in the sequence cannot be used to predict, using any statistical method, the values of subsequent members of the sequence. (Clearly, the deterministic formula used to generate the pseudo-random number sequence can be used to predict values of the sequence. However, in a statistical test this formula would not be "known" *a priori.*)

For any sequence $s_1, s_2, s_3, \ldots, s_{n+N+1}$ it is possible to find an **autoregression model of order** $n \leq N + 1$ of the form

$$s_i = a_i s_{i-1} + a_2 s_{i-2} + \cdots + a_n s_{1-n} + \varepsilon_i, i > n + N + 1, \tag{A.2}$$

where $a_1, a_2, \ldots, a_n$ are **autoregression coefficients** and ε_i is the error in predicting s_i. The procedure for obtaining the values of the n autoregression coefficients from $n+N+1$ members of a sequence is given in Section A.4. Equation (A.2) uses only *a posteriori* knowledge of a subsequence of $n+N+1$ members of a sequence to predict, on the basis of statistics, the value of the ith member of the sequence, where $i \geq n+N+2$. The accuracy with which Eq. (A.2) can be used to predict the ith member of the sequence also depends on the degree to which the sequence represents a deterministic process; that is, the degree to which the ith member is correlated with the previous $n+N+1$ members. If the error ε_i is sufficiently large, the ith member can be said to be uncorrelated with the previous $n+N+1$ members of the sequence. This, of course, is the desired property of a good pseudo-random number generator.

Equation (A.2) can be used to generate a subsequence of numbers whose values can then be compared, using the standard χ^2test, with the corresponding actual subsequence produced by the pseudo-random number generator. If the pseudo-random number sequence is uncorrelated, we should expect to obtain a large value for χ^2.

A.2.3 Property 3

The period of the sequence of random numbers should be **sufficiently long** to complete the simulation without repeating any part of the sequence. The period of a pseudo-random number generator is the count of numbers that can be generated before the sequence is repeated. Every pseudo-random number generator has a period.

A.2.4 Property 4

The PRNG should be **repeatable**. That is, given the same starting values, called **seeds**, the generator should produce the same sequence of numbers. This property is essential in debugging Monte Carlo algorithms.

A.2.5 Property 5

The PRNG should provide **long, disjoint sequences**. This means that different users performing subparts of a same overall simulation should be able to work simultaneously without danger of inadvertently using the same sequence of random numbers, thereby possibly introducing a bias to the overall simulation. If a PRNG does not have this property then each part of the simulation must be carried out serially. That is, the first part must be carried out and the final pseudo-random number recorded so that this number can be used as the seed to start the PRNG for the second part of the simulation, and so forth.

A.2.6 Property 6

The PRNG should be **portable**. Many "random" number generators written in FORTRAN use processor-specific programming shortcuts that do not work on all processors.

A.2.7 Property 7

The PRNG should be **efficient** because billions of random numbers may be needed in modern Monte Carlo simulations. Trade-offs will usually be required between efficiency and achievement of the other six properties of a "good" PRNG, however.

The statistical properties of a relatively short sequence of members of a long random number sequence are generally less important than the statistical properties of the larger sequence. However, if the MCRT method is to be used to model a system so large that only a few hundred energy bundles can be emitted per surface, the statistical properties of relatively short sequences will be important.

A.3 A "Minimal Standard" Pseudo-Random Number Generator

Park and Miller [3] studied the manifold of available PRNGs and concluded that the family of **multiplicative linear congruential generator** (MLCG) algorithms first proposed by Lehmer [4] can form the basis for what they refer to as a "minimal standard" PRNG. At the center of the MLCG algorithm is the recursion formula

$$s_{i+1} = (as_i + c) \bmod m, \tag{A.3}$$

where s_{i+1} and s_i are successive members of the pseudo-random sequence $s_0, s_1, s_2, \ldots$. In Eq. (A.3) the coefficient a is a well-chosen integer multiplier in the range $[2, m-1]$, the constant c is a number that can be (and usually is) equal to zero, and m is an integer usually equal to or slightly smaller than the largest integer that can be represented in one computer word. The notation "n mod m" means "the modulus of n/m," where the modulus operator produces the first remainder of the division of two integers. For example,

$$13 \bmod 15 \equiv \text{first remainder of the operation} \quad 15\overline{)13}^{\,0} = 13 \qquad \begin{array}{r} 13 \\ \underline{0} \\ 13 \end{array} \tag{A.4}$$

and

$$18 \bmod 15 \equiv \text{first remainder of the operation} \quad 15\overline{)18}^{\,1} = 3. \qquad \begin{array}{r} 18 \\ \underline{15} \\ 3 \end{array} \tag{A.5}$$

The initial value of the sequence, s_0, is called the **seed** and must be provided by the user. Most program languages, including Matlab, have built-in modulus functions.

The members of the sequence produced by Eq. (A.3) are integers whose values range between 1 and $m - 1$. The s sequence is converted into a sequence of pseudo-random numbers R, uniformly distributed between zero and unity, by dividing by m. Note that neither zero nor unity can result from this and similar MLCG algorithms with the constant c equal to zero. However, if the resulting sequence contains billions of members, each expressed to 32-bit precision, the difference between zero and unity, respectively, and the smallest and largest members is arbitrarily small. A "good" PRNG is one for which each seed value will lead to a different random number sequence of the same (known) fixed period. As an example, consider the MLCG algorithm $s_{i+1} = 6s_i \bmod 13$. Then if the seed s_0 is taken to be 1, the resulting sequence is 1, 6, 10, 8, 9, 2, 12, 7, 3, 5, 4, 11, 1, … and the sequence repeats itself. It is clear in this example that the period is short and that the number of significant figures in the corresponding random number sequence ($R_i = s_i/13$, $i = 1, 2, 3, \ldots, 12$) is small. This suggests that in a practical PRNG, the quantities a and m should be as large as possible (although not all values of $a < m$ lead to **full-period sequences**; that is, to sequences whose period is $m - 1$).

Down through the years many MLCG algorithms have been proposed and implemented as "commercial release" PRNGs. Subsequent statistical studies have revealed that some are quite "good" as defined in Section A.2 while others are surprisingly "bad." Some combinations of values of a and m are known to be better than others in their ability to produce long-period sequences of statistically random numbers.

The largest **prime** integer that can be represented on a 32-bit machine is $2^{31} - 1$, and so this number has emerged, because of number-theory considerations beyond the scope of this book, as the favorite choice for m in 32-bit operating systems. Research into the best multipliers has led many to conclude that $a = 7^5 = 16\,807$ is a good choice. The resulting MLCG algorithm,

$$s_{i+1} = 16{,}807 s_i \bmod (2^{31}-1) \tag{A.6}$$

is a full-period algorithm. Of practical concern is that 46 bits would be required to store the largest possible value of the product $16\,807 s_i$. The potential for integer overflow therefore requires careful programming if an MLCG algorithm based on these values of a and m is to be universally portable.

A double-precision FORTRAN version of an MLCG algorithm, RANMAR, by Marsaglia and Zaman [5], based on Eq. (A.6), meets all of the requirements laid out in Section A.2. For example, it is capable of creating 900 million different random number sequences with each subsequence having a period of approximately 10^{30}!

RANMAR actually consists of two subroutines. Before the first call to RANMAR, the FORTRAN command CALL RMARIN(IJ, KL) must be executed, where the integer variables IJ and KL (the seeds) have already been defined such that $0 \leq \text{IJ} \leq 31\,328$ and $0 \leq \text{KL} \leq 30\,801$. The second subroutine, RANMAR, is the actual pseudo-random number generator. Then each time a random number is needed, the FORTRAN command R = RANMAR() is executed, where R contains the new random number after execution of the command. We, and presumably others, have translated RANMAR for use with Matlab.

RANMAR is structured so that a large number of teams working independently at different locations can perform parts of the same Monte Carlo simulation without danger of a given random number subsequence appearing in any two parts. That is accomplished by assigning each team a different (randomly assigned) value of the seed IJ and then allowing each team to pick its own value of the second seed KL.

A.4 Autoregression Analysis

Most physical processes can be represented by a deterministically based time series.[1] For example, the distance $x - x_0$ traveled in time t by an

[1] It is usual to refer to any equally spaced series of observations as a "time" series even if the independent variable is, say, space rather than time.

object dropped from rest in a vacuum is described by

$$x - x_0 = \frac{gt^2}{2}, \tag{A.7}$$

where g is the acceleration of gravity. Often the process in question is much more complex, being governed by a large system of coupled equations, some of which may be nonlinear. A well-known example is that of an aircraft maneuvering in a combat situation. In this case, while the aircraft trajectory is governed by a set of equations, a gunner tasked to shoot down the aircraft is neither privy to the equations nor does he have time to solve them. The pilot often renders the gunner's job even more difficult by executing random twists and turns. How might the gunner succeed in shooting down the aircraft?

When first faced with the fire-control problem described above, early designers of radar-aimed anti-aircraft guns developed a technique called Kalman filtering, or **autoregression analysis**. The underlying idea is that the gunner does not need to know or solve the actual set of equations governing the trajectory of the aircraft if he can find a simpler model that represents the motion of the target over a short but sufficiently long period of time. The assumption in autoregression analysis is that information gained by observation of the trajectory over a limited period of time can be used to predict the trajectory for a brief period of time into the future.

Suppose the observed trajectory is known in terms of a **sampled time series**, as illustrated (in one dimension) in Figure A.1. In the figure the filled symbols represent the actual sequence of positions making

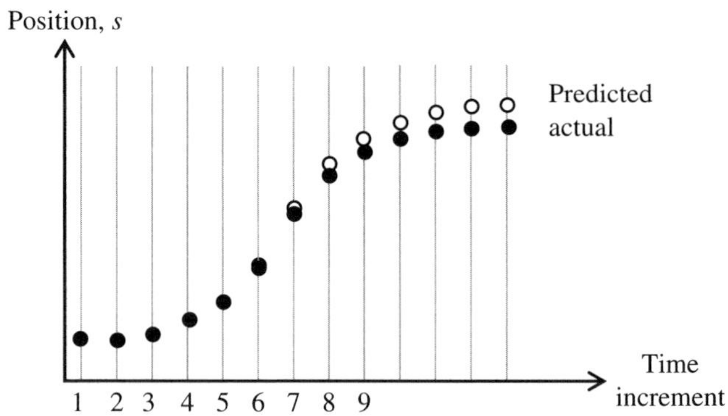

Figure A.1 A hypothetical sampled time series representing a segment of the trajectory followed by a maneuvering aircraft.

up the trajectory, and the open symbols represent positions predicted using a fourth-order autoregression model. Then the open symbol at time increment 9 represents the position s'_9 predicted by a fourth-order autoregression model of the form

$$s'_9 = s_9 - \varepsilon_9 = a_1 s_8 + a_2 s_7 + a_3 s_6 + a_4 s_5. \tag{A.8}$$

In Eq. (A.8), ε_9 is the error between the observed and predicted positions at the ninth time increment; that is,

$$\varepsilon_9 = s_9 - s'_9, \tag{A.9}$$

and the coefficients a_1, a_2, a_3, and a_4 are the **autoregression coefficients** that correlate the position at the ninth time increment with the positions at the previous four time increments. We will learn in Section A.4.1 that the order n of the autoregression model is limited by the number of observations available for computing the n autoregression coefficients. The general form of Eq. (A.8) for an nth-order autoregression is given by Eq. (A.2).

The autoregression coefficients can be thought of as sensitivities, or slopes, in a linear regression, as illustrated in Figure A.2 for the case of a second-order autoregression model. In this simple example it is clear that

$$s'_3 = s_3 - \varepsilon_3 = \frac{\partial s_3}{\partial s_2} s_2 + \frac{\partial s_3}{\partial s_1} s_1 = a_1 s_2 + a_2 s_1. \tag{A.10}$$

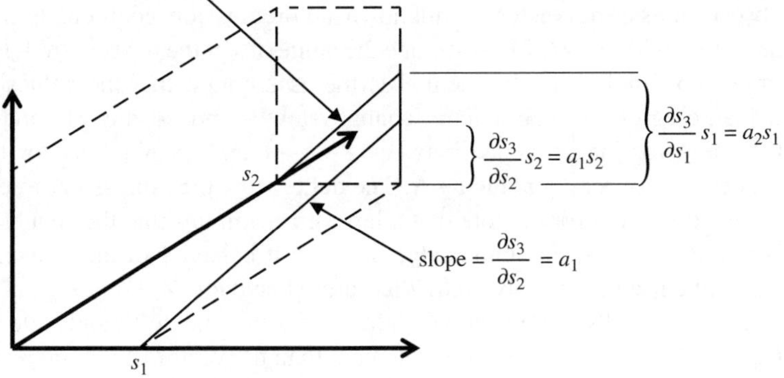

Figure A.2 Illustration of the interpretation of the autoregression coefficients as slopes for a second-order autoregression.

Since the regression is linear, the error $\varepsilon_3 = s_3 - s_3'$ will depend on the size of the sampling interval $\Delta s = s_3 - s_2 = s_2 - s_1$ and on the curvature of the true trajectory in the vicinity of the point s_3. Therefore, it is not always the case that an autoregression of the highest available order n will yield the smallest error. In the current example we may say that a_2 is the sensitivity of $s_3$to s_1, all other variables being held constant. Figure A.2 illustrates a situation in which s_3 is more sensitive to s_1 than to s_2.

A.4.1 Calculation of the Autoregression Coefficients

We consider a sampled time series such as the one illustrated in Figure A.1 and assume that there exist n autoregression coefficients a_1, a_2, …, a_n such that, for the range of discrete observations of s between $i-n$ and $i+N$,

$$\begin{bmatrix} s_i \\ s_{i+1} \\ \cdots \\ s_{i+N} \end{bmatrix} = \left[\begin{bmatrix} s_{i-1} & s_{i-2} & \cdots & s_{i-n} \\ s_i & s_{i-1} & \cdots & s_{i+2-n} \\ \cdots & \cdots & \cdots & \cdots \\ s_{i+N-1} & s_{i+N-2} & \cdots & s_{i+N-n} \end{bmatrix}\right] \begin{bmatrix} a_1 \\ a_2 \\ \cdots \\ a_n \end{bmatrix}, \tag{A.11}$$

or

$$[S] = [[\overline{\overline{S}}]][A]. \tag{A.11a}$$

The form of Eq. (A.11) assumes the availability of a sequence of $n+N+1$ observations. Equation (A.11) is solved to obtain the n coefficients of an nth-order autoregression model based on the $n+N+1$ observations. The vector of unknown autoregression coefficients a is then used with Eq. (A.2) to estimate the value of s at the next $(i+N+1\text{st})$ time step. Implicit in the method is the assumption that the values of the autoregression coefficients change slowly and so are essentially the same for the sequence between $i-n+1$ and $i+N+1$ as for the sequence between $i-n$ and $i+N$. The degree to which this is not true is reflected in the error ε. Note that a large error implies that the sequence is random; that is, that knowledge of the past behavior of the sequence cannot be used to reliably predict its future behavior.

The vector [**S**] of observed values of s on the left-hand side of Eq. (A.11) has at least one more element than the vector [**A**] of unknown autoregression coefficients on the right-hand side. As a consequence of this, the matrix $[[\overline{\overline{S}}]]$ is not square. Therefore, to solve for the vector of unknown autoregression coefficients, we must begin by multiplying

Eq. (A.11) by the transpose of $[[\bar{\bar{S}}]]$,

$$[[\bar{\bar{S}}]]^t[S] = [[\bar{\bar{S}}]]^t [[\bar{\bar{S}}]][A], \tag{A.12}$$

or

$$\begin{bmatrix} \sum_{j=i}^{i+N} s_j s_{j-1} \\ \sum_{j=i}^{i+N} s_j s_{j-2} \\ \cdots \\ \sum_{j=i}^{i+N} s_j s_{j-n} \end{bmatrix} = \left[\left[\begin{matrix} \sum_{j=i}^{i+N} s_{j-1}s_{j-1} & \sum_{j=i}^{i+N} s_{j-1}s_{j-2} & \cdots & \sum_{j=i}^{i+N} s_{j-1}s_{j-n} \\ \sum_{j=i}^{i+N} s_{j-2}s_{j-1} & \sum_{j=i}^{i+N} s_{j-2}s_{j-2} & \cdots & \sum_{j=i}^{i+N} s_{j-2}s_{j-n} \\ \cdots & \cdots & \cdots & \cdots \\ \sum_{j=i}^{i+N} s_{j-n}s_{j-1} & \sum_{j=i}^{i+N} s_{j-n}s_{j-2} & \cdots & \sum_{j=i}^{i+N} s_{j-n}s_{j-n} \end{matrix}\right]\right] \begin{bmatrix} a_1 \\ a_2 \\ \cdots \\ a_n \end{bmatrix}. \tag{A.13}$$

We are now able to solve Eq. (A.13) directly for the vector of autoregression coefficients,

$$[A] = [[\bar{\bar{\Sigma}}]]^{-1} \ [\Sigma], \tag{A.14}$$

where $[[\bar{\bar{\Sigma}}]]^{-1}$ is the inverse of the matrix in Eq. (A.13) and $[\boldsymbol{\Sigma}]$ is the vector on the left-hand side.

A.4.2 Example

We now consider Eq. (A.7), the model for the distance an object falls from rest in a vacuum. Table A.1 lists the "observed" vertical displacement of the object, based on Eq. (A.7), at 11 equally spaced time steps beginning at $t = 0\,s$. We wish to predict the displacement 7, 8, 9, and 10 s after the object is released using a third-order autoregression model with $N = 3$. Thus, we will need to use the first seven observed displacements to compute the three autoregression coefficients. The results are given in Table A.1 and in Figure A.3.

Inspection of Table A.1 reveals that the error in the predicted displacement is less than 1% at the eighth and ninth time steps (t = 7 and 8 s), but then doubles to 2 and 4% at the ninth and tenth time steps. From this we may conclude that the model performance degrades as the model is applied farther from the time interval used to compute the autoregression coefficients. In fact, this is a general property of autoregression models.

Table A.1 Vertical displacement observed and predicted from a third-order autoregression model based on $n+N+1=7$ observations ($\alpha_1 = 2.9885$, $\alpha_2 = -2.9591$, $\alpha_3 = 0.9553$).

Time (s)	Displacement observed (m)	Displacement predicted (m)	$\frac{\text{Observed} - \text{Predicted}}{\text{Observed}}$
0	0	—	—
1	4.9	—	—
2	19.6	—	—
3	44.1	—	—
4	78.4	—	—
5	122.6	—	—
6	176.5	—	—
7	240.3	239.6	0.0029
8	313.8	310.8	0.0096
9	397.1	388.6	0.0214
10	490.3	470.3	0.0408

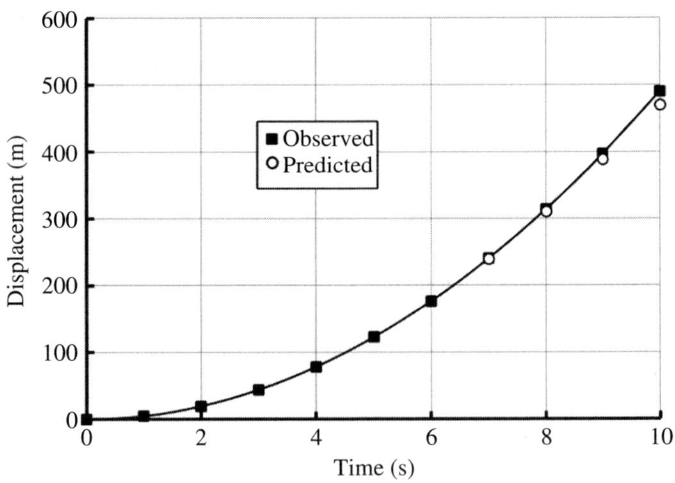

Figure A.3 Observed and predicted displacements using a third-order autoregression model for an object dropped from rest falling under the influence of gravity in a vacuum.

Problems

A.1 A PRNG algorithm has been proposed in which an n-digit integer, where n is an even number, is squared and then the middle n digits of the $2n$-digit result are isolated and squared, and so forth. For example, if $n = 4$ and the integer $s_0 = 1234$ is selected as the seed, then the first few members of the resulting sequence are 1234, 5227, 3215, 3362, 3030, 1809, 2724, 4201, 6484, 422, 1780, 1684, ... Dividing each of these by the maximum possible value, 9999, yields the pseudo-random number sequence (rounded to eight significant figures) 0.12341234, 0.52275227, 0.32153215, 0.33623362, 0.30303030, 0.18091809, 0.27242724, 0.42014201, 0.64846484, 0.04220422, 0.17801780, 0.16841684. Write a computer program to create a sequence of pseudo-random numbers based on this algorithm. Test the program for the case of $n = 8$ and $s_0 = 12\,345\,678$.

A.2 Explore the capabilities of the PRNG algorithm created in Problem A.1. Specifically, determine how the length of the sequences generated depends on the values of n and s_0. Discuss the behavior of the PRNG for $n = 2$, 4, 6, and 8 with at least three values of the seed for each value of n.

A.3 Run the PRNG in Problem A.1 with $n = 8$ and $s_0 = 12345678$ until a sequence of 100 (or the largest number of random numbers available if less than 100) random numbers has been obtained. Remember to scale the sequence by dividing by 99 999 999 so that the values vary between zero and unity.

(a) Use the χ^2 test to evaluate the uniformity of distribution of the sequence obtained. (Use 10 bins.)

(b) Use a fourth-order autoregression model to test the degree of correlation of the sequence.

A.4 Write a computer program that implements an MLCG algorithm based on

$$s_{i+1} = 5s_i \bmod 101 \tag{A.15}$$

to produce a sequence of pseudo-random numbers, and execute the program with $s_0 = 7$. What is the period of the random number sequence produced by this algorithm? Is the resulting sequence a full-period sequence?

A.5 For the pseudo-random number sequence produced by the algorithm in Problem A.4, use the χ^2 test to evaluate the uniformity of distribution of the sequence obtained. (Use 10 bins.)

A.6 Use a fifth-order autoregression model to test the degree of correlation of the pseudo-random number sequence produced by the algorithm in Problem A.4.

A.7 Verify the autoregression coefficients (given in the title of Table A.1) for a third-order autoregression model of a body falling under the influence of gravity (Eq. A.7) using the first seven "observed" displacements in Table A.1. Note that $i = n = N = 3$ in this case. Then use the autoregression model to verify the predicted displacements for $t = 7$, 8, 9, and 10 s given in Table A.1. Also compute the percent difference between the observed and predicted displacements at these times.

A.8 Compute the four autoregression coefficients for a fourth-order autoregression model of a body falling under the influence of gravity (Eq. A.7) using the nine "observed" displacements in Table A.1. Note that $i = n = N = 4$ in this case. Then use the autoregression model to verify the predicted displacements for $t = 9$ and 10 s given in Table A.1. Also compute the percentage difference between the observed and predicted displacement at these times. Are the percentage differences smaller for the fourth-order autoregression model than they were for the third-order model considered in Section A.4.2?

References

1. James, F. (1990). A review of pseudo-random number generators. *Computer Physics Communications* 60: 329–344.
2. Dally, J.W., Riley, W.F., and McConnell, K.G. (1984). *Instrumentation for Engineering Measurements*, 541. New York: Wiley.
3. Park, S.K. and Miller, K.W. (October 1988). Random number generators: good ones are hard to find. *Computing Practices* 31 (10): 1192–1201.
4. Lehmer, D.H. (1951). Mathematical methods in large-scale computing units. *Annual Computing Lab, Harvard University* 26: 141–146.
5. Marsaglia, G., and Zaman, A., (1987). Toward a Universal Random Number Generator. Report FSU-SCRI-87-50, Florida State University.

Index

The Monte Carlo Ray-Trace Method in Radiation Heat Transfer and Applied Optics,
First Edition. J. Robert Mahan.

This Work is a co-publication between John Wiley & Sons Ltd and ASME Press.
Companion website: www.wiley.com/go/mahan/monte-carlo